Textbook on Weed Science

NIPA® GENX ELECTRONIC RESOURCES & SOLUTIONS P. LTD.
New Delhi-110 034

About the Author

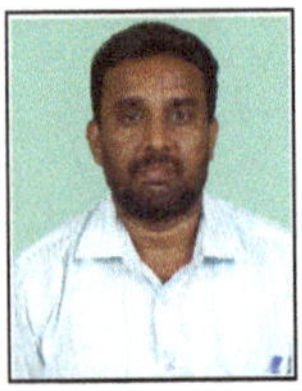

Dr. A. Veeramani, Professor is presently working at Agricultural College and Research Institute, Tamil Nadu Agricultural University, Madurai, Tamil Nadu, India. The author has been awarded with M.Sc., (Ag.) and Ph.D in Agronomy at TamilNadu Agricultural University during 1995 and 2004 respectively. He has served in the cadres of Assistant Professor, Associate Professor and Professor in the discipline of Agronomy at various Research stations and Teaching campuses of Tamil Nadu Agricultural University for more than 21 years besides serving in the capacity of Associate Professor and Head of the Research Station and Programme Co-ordinator of KVK in Tamil Nadu. During that period he has made various research achievements in semi-dry rice, rice-fallow crops and weed control in various crops. Under teaching, he taught several UG, PG and Ph.D courses in Agronomy. He has guided four PG and a Ph.D student in Agronomy. With regard to publications, there were 4 National standard reference books written by him in the areas of principles, terminologies, irrigation and weed control. In addition, 5 research papers in international and 11 in National level journals and 41 popular articles published besides 32 research papers in national and international seminar and conferences.

Participated and presented a research paper on "Motorized seed drill for Rice-fallow pulse" in "The IRES 292nd International Conference on Food and Agricultural Engineering" at Chicago, USA during 24th and 25th Nov, 2017. A machine "Rice fallow pulse planter" was also developed by him through an externally funded project by DST, New Delhi. He was also given a distinction with "The IRES Excellent paper award" in the above said conference at Chicago, USA besides "Fellow of Hind Agri-Horticultural Society (FHAS)".

Textbook on Weed Science
Principles & Practices

A. Veeramani
Professor
Agricultural College and Research Institute
Tamil Nadu Agricultural University
Madurai, Tamil Nadu

NIPA® GENX ELECTRONIC RESOURCES & SOLUTIONS P. LTD.
New Delhi-110 034

NIPA® GENX ELECTRONIC
RESOURCES & SOLUTIONS P. LTD.

101,103, Vikas Surya Plaza, CU Block
L.S.C. Market, Pitam Pura, New Delhi-110 034
Ph : +91 11 27341616, 27341717, 27341718
E-mail: newindiapublishingagency@gmail.com
web: www.nipabooks.com

For customer assistance, please contact
Phone: + 91-11-27 34 17 17
Fax: + 91-11- 27 34 16 16
E-Mail: feedbacks@nipabooks.com

ISBN: 978-93-58870-65-7

Composed and Designed by NIPA®.

TAMIL NADU AGRICULTURAL UNIVERSITY
Coimbatore - 641 003
Tamil Nadu, India

Prof. N. Kumar
Ph.D., F.H.S.I.
Vice-Chancellor

Foreword

There are several factors which cause loss in agriculture. However, overall annual loss of agricultural produce due to weeds is higher which accounts for 45%. The total annual economic loss amounts to Rs. 20-28 billion in the country. Under the above situation, to achieve better weed control, proper use of available techniques at field level is a must. It is possible by having thorough knowledge on wee biology and weed ecology, as this underlines the basic idea about weeds that are infesting crop fields. Taking into consideration of these facts, this book is written duly incorporating needy measures required for weed control in various crops including non-cropped areas like roads sides, barren lands, waster lands, school ground, public parks and water bodies that face weed menace.

Of the available weed control measures, use of herbicides is found to be the most effective and economic weed control method. Moreover, this is the best alternate method under labour scarce situation which is not unusual in the present day content in almost all the countries in world. Even though herbicides posses several advantages in terms of economics and higher weed control efficiency as compared to rest of the methods, it also causes ill effects which include reduction in population of beneficial microorganism in the soil, poor soil health due to its residual effects and environmental hazards due to indiscriminate use of herbicides irrespective of crops. For instance, indiscriminate use of Isoproturon in wheat imparts herbicide resistance in *Phalaris minor* and continuous use of Butachlor in rice leads to herbicide resistance in *Echinochloa colona.* Taking into account of the same, several strategies were developed in the recent past by following holistic approach which encompasses the development of herbicide resistant crops through biotechnological approaches, use of herbicide mixtures, robotic weed control and real - time weed control which would certainly reduce the ill-effects in soil and crops and keep the environment pollution free besides giving better weed control.

This book provides required information on various aspects related to weed management which includes principles as well as practices for application in cultivated lands and un-cropped area. I am sure that this book will be highly useful to the students, scientist and farmers who are interested in weeding out the weeds. I complement the author for his effort.

N. Kumar

Preface

There are several factors which involve in influencing the growth and development and yield of crops cultivated in the fields. On the other hand, these factors also affect crop growth. However, under biological factors, weed is one which is said to be harmful in affecting the growth of the crops due to its high degree of competitive nature for resources like moisture, light, nutrients, space etc. beside more menace due to its interference in human activities in crop fields. Even though, there are various losses caused by biological factors, the loss of yield due to weeds seemed to be very high which accounts for >50% in severe cases compare to pests and diseases which are proven to be next major factors and yield loss due to them accounts for 18% and 16% respectively. Hence, utmost care in controlling weeds in the cultivated lands is imperative as loss in agricultural production due to weeds is estimated to be 25% in least developed countries. For achieve better management of weeds irrespective of area that are infested with weeds, it necessarily want of high tech measure under weed control. Taking into account of the above, this book was prepared highlighting the strategies made available in several sources.

This book was incorporated with general information about weeds with need based data on losses caused by weeds in agriculture and allied fields which include livestock, horticulture crops etc. and also national level loss in terms of reduction in food grain production in Indian economy. As management of weeds being a core area in crop production, it was dealt in detail on weed biology, weed ecology and also weed survey including sampling techniques. This was followed by various weed control methods specific to crops belonging to agriculture and horticulture. Of the weed management practices, integrated weed management was also discussed as well highlighting important crops grown in India as to follow the same by the farmers and extension workers for recommendation to the needy people. This book also comprises of herbicide history and its classification and characteristics with suitable examples. In addition, mode of action of herbicides, herbicide mixture, combinations, rotations, and their interaction were also given as separate chapter. In the recent decade, development of herbicide resistant crops are getting momentum world over, considering its potentiality, information on the same was also given in detail with suitable findings. Weed shift and its management and economic analysis of weed management were another chapters included in this book.

I hope that the contents incorporated in this book relevant to weed management would certainly help improve the knowledge of the people who dealt with said area. The author express his gratitude and sincere thanks to the Dean, Agricultural College and Research Institute, Kudumiyanmalai and his colleagues for having extended their constant encouragement and valuable suggestions which made this book into constructive in serving better knowledge on weed science.

December, 2018 **A. Veeramani**

Acknowledgement

I reproduced in this book certain material which includes figures, and discussions on them, published in other books and journals. I hereby acknowledge the authors and scientists for having contributed their sources like reference books and reports which I availed as below :

1. Weed Control Unit, TNAU, Coimbatore, India: To reproduce: (a) in Chapter 1.1 (Fig.1.1.7, 1.1.9, 1.1.12, 1.1.13, 1.1.14, 1.1.15, 1.1.16, 1.1.18, 1.1.21, 1.1.22, 1.1.23, 1.1.24, 1.1.38, 1.1.41 and 1.1.42.) of this edition.
2. AICRP-WM in www.dwr.org : To reproduce in Chapter 4.0 (Table 4.1, 4.2) of this edition. Table 4.1 'Cost of cultivation of Onion (Rs./ha)' and Table 4.2 'Partial Budget showing change between users and non-users of Herbicide in cultivating Maize'.
3. Kalyani Publishers, New Delhi, India: To reproduce and adapt, in Chapter 2.6 of this edition, Fig.2.6.1 'Schematic presentation of the electron transport and phosphorylation reactions considered to occur in chloroplast lamella' published, in 2014, in a book entitled 'WEED MANAGEMENT', fourth revised edition, authored by U.S.Walia.
4. Weed Science Society of America, Lawrence, KS, USA: To reproduce the following figure: (a) In Chapter 2.6 of this edition: Figure 2.6.2 'The triazine block electron in PS II by displacing QB from its binding site on the D_1 protein' from a chapter on 'Interactions of herbicides with photosynthetic electron transport', authored by Furest, F.P. and M.A.Norman and published, in 1991, in WEED SCIENCE, Vol.39.
5. Kalyani Publishers, New Delhi, India: To reproduce the following figures: (a) In Chapter 2.9 (Fig.2.9.2-2.9.3) of this edition: Figure 2.9.2. 'Gene splicing in recombinant DNA' and Figure 2.9.3. 'Plant genetic engineering' from a Chapter on 'What is new in the Weed Management system', authored by S. Subramaniam, A. Mohamed Ali, P. Muthukrishnan, R. Jaya Kumar, B. Chandrasekaran and S.Vallal Kannan and published, in 2011, in a book entitled 'All About Weed Control'.

A. Veeramani

Contents

Unit-1: Weeds

List of Colour Plates

Unit-1: Weeds

1.1

Introduction to Weeds

The association of weeds with field crops was started when ancient man started cultivating plants as crops commercially. However, type of weed flora went on shifting with the passage of time, due to changing climatic conditions, cropping pattern, agronomic practices, crop cultivars etc. When crops are cultivated invariably a variety of volunteer undesirable vegetation come up simultaneously which competes with the crop for all growth factors *viz.*, nutrients, soil moisture light, space etc.

Definition of Weed

Weed is a plant that originates under a natural environment and in response to imposed and natural environments, evolved and continues to do so as an interfering associate with our desired plant and activities.

Weed is a plant growing out of place. They are unwanted not useful, persistent and prolific, effectively competing with the beneficial and desirable crop plants for space, nutrients, sunlight, and water, interfere with agricultural operations and thereby reducing the yield and quality of produce.

Weeds are unwanted and undeserved plants that interfere with the utilization of the land and water resources and thus adversely affect crop production and human welfare. **Thus a plant out of its place or a plant growing where it is not desired is a weed. This** definition was given by Buchholtz in 1967. *E.g.* Bajra is a weed in pulse, pulse is a weed in bajra, tomato is a weed in brinjal field. **Weed** is defined as the unwanted, undesirable plant growing out of their proper place which interfere with the utilization of natural resources, prolific, persistent, competitive, harmful and even poisonous in nature and can grow in adverse climatic conditions (Jethro Tull: Father of Weed Science).

Weediness: Weediness is defined as the state or condition of a field, flower beds, lawns and so forth in which there is an abundance of weeds.

Weeds are plants that are unwanted in a given situation and may be harmful, dangerous or economically detrimental. Weeds are a serious threat to primary production and biodiversity. They reduce farm and forest productivity, displace native species and contribute significantly to land and water degradation. The costs of weeds to the natural environment are also high, with weed invasion being ranked second only to habitat loss in causing biodiversity decline.

Weed can also be defined as "a plant which grows voluntarily at places where it is not wanted and grows at places where other useful plants grow". Plants are also considered weeds "whose virtues are not known yet." Some weed plants are very useful e.g. *Datura stramonium* and *Solanum xanthocarpum* (sankanmoli) as these have medicinal value but are still termed as weeds. Plants are considered weeds "which interfere with the utilization of land and water resources and hamper the activities of human beings directly or indirectly".

Characteristics of Weeds

Like crop plants, weeds also possess some features by virtue of which these are recognised as wild plants or unwanted plants. A knowledge about these features will help in developing suitable methods for their control by studying their most sensitive stage in their life cycle is very important. Moreover, knowing about the characteristics of these weeds will help in studying means of natural adaptations as well as extent of losses which these weeds can render to human beings and crops are also important. The following features (characteristics) are common for most of the weeds growing in association with crops or places other than the cultivated lands.

Can Tolerate Adverse Climatic Conditions

Weeds have the capacity to thrive under adverse climatic conditions under which crops cannot be grown successfully. This is because of better adaptation features of weed plants to establish under adverse/odd type of situations/habitats. Moreover, being wild plants their genetic constitution due to natural selection is such which can resist/tolerate adverse climatic conditions. Due to better adaptability to a particular place/region, weeds get the chance to suppress the crop(s) and even grows luxuriantly under such conditions. For examples, *Calotropis procera* (Akk) and *Saccharum spontaneum* (Kahi) can tolerate high temperature as well as limited soil moisture conditions due to thick cuticle and deep roots. Similarly, *Carthamus oxyacantha* (Pohli) and *Asphodelus tenuifolius* (Piazi) grow well in a habitat having limited available soil moisture and even over power the crop in which these weeds grow. Some annual weeds produce more than one seed crop per year and seed is produced as long as growing conditions permit. Each generation is capable of producing large number of seeds per plant and seed is produced over a wide range of environmental conditions.

Competitive and Aggressive in Nature

Weeds which generally grow in close vicinity of the crop plants, rob the crops for various inputs which are provided for their growth and development. Since both weeds and crops are plants, have basically the same requirements for inputs for normal growth and development. When weeds with taller and dense canopies grow in close proximity with the crop, these suppress the growth of crop plants adversely. Under such a plant community, more competition exists among the different plant species. Moreover, weeds being wild, always have an edge over the crop plants with respect to their growth and developments because of early germination and quick initial growing habits. The presence of *Trianthema portulacastrum* in maize, cotton, sugarcane, pulses, potato etc. is the true examples of competitiveness and aggressiveness and thus this weed dominates these crops during initial growth stages. The aggressiveness of a weed species is associated with its growth habits such as tall growing, more leaf area, rapid early growth, spreading habit and efficient root system. Most of the weeds including *Phalaris minor*, *Avena ludoviciana*, *Echinochloa crusgalli*, *Eleusine indica*, *Digitaria sanguinalis*, *Cyperus rotundus*, *Celosia argentea*, *Sorghum halepense*, *Cynodon dactylon* etc. being aggressive, competes with the crops and result in marked reduction in yield of these crops. Weeds have great competitive ability for nutrients, light and water and can compete by special means, (e.g. Rosette formation, climbing growth and allelopathy). Weeds are ubiquitous (present everywhere). They exist everywhere, where we practice agriculture.

Resist Control/Eradication

Weeds due to the presence of special structures such as spines or thorns, hinder their removal and hence continue growing with the crop plants till harvest. The presence of spines in *Carthamus oxyacantha* (Pohli) and *Cirsium arvense* (Leh/Canada thistle) hinder their removal from wheat. Other weeds such as *Amaranthus spinosus* (Kandyali Chulai), *Cenchrus catharticus* (Lendhra), *Tribulus terrestris* (Bhakhra), *Zizyphus rotundifolia, Rumex spinosus* etc. resist removal under different situations due to special features i.e. spines/thorns. Perennial weeds such as *Cynodon dactylon*, *Sorghum halepense* etc. are able to regenerate from their underground (vegetative) part(s) and hence continue their infestation with crops even if the aerial parts (above ground) are removed before seed setting. Similarly tubers of *Cyperus rotundus* cannot be removed with hand hoeings or hand pullings, as these are very deeply placed in the soil. Many weeds have adaptation mechanism that repels grazing, such as spines, odd taste or odour. Weeds resist control, including herbicides.

Morphological Similarities

Some weeds resemble morphologically with the associated crop and it is very difficult to identify these especially at the time of hoeing. For example, *Phalaris minor* (Gullidanda /sitti) and *Avena ludoviciana* (wild oats/jangli javi) resemble with wheat plants particularly during initial stages of crop growth. So, at the time of hand hoeing, it becomes impossible / to remove such type of weeds thoroughly. Similarly plants of swank (*Echinochloa crusgalli*) resemble with rice seedlings both in nursery and in transplanted rice crop. Its removal from rice nursery is very difficult due to morphological similarities and these weed plants are frequently transplanted in field along with rice seedlings. This feature of weeds, makes weed control methods as ineffective and results in enrichment of weed seed bank year after year.

Mimic Weeds

High Reproductive Capacity

Weeds are prolific in nature and have very high reproductive potential and produce huge quantity of seeds in a short spell of time and hence pose serious problem in cultivation of crops. In other words weeds deposit a large quantity of seeds to the seed bank of soil and create trouble since these seeds are the means by which weeds are introduced and spread in the field during succeeding years. A single plant of *Chenopodium album* in wheat produces 10,000 to 15,000 seeds and one plant of *Phalaris minor* produces 3,000 to 5,000 seeds in a single season. Similarly, an individual plant of *Trianthema portulacastrum* when growing without any competition produces lacs of seeds.

The seed production capacity of some of the weeds is as follows:

Ontogeny	Seeds/plant	Name of weed/crop	Seeds/plant
Perennials	16,629	*Amaranthus retroflexus*	1,96,405
Biennials	26,600	*Solanum nigrum*	1,78,000
Annuals	20,832	*Chenopodium album*	72,000
		Trianthema portulacastrum	52,000
		Wheat & Rice	90 to 100

A few weeds may produce seed through apomixis i.e. without fertilization. Eg. Ferns reproduce by spores.

Persistent in Nature

Germination is retained for a longer period of time even for 10 to 20 years after their burial in soil. The seeds of *Chenopodium album* can germinate even upto 20-25 years, *Convolvulus arvensis* up to 20 years, *Trianthema portulacastrum* for 15 years and *Phalaris minor* from 4 to 5 years after burial in the soil. The seed dormancy plays an important role in the continuous infestation of a weed in a particular crop. Likewise most of the other weed seeds remain dormant for an extended period. In few cases weed seeds retain their viability even after passing through the digestive tract of the animals e.g. *Trianthema portulacastrum*, *Chenopodium album, Eleusine indica, Phalaris minor, Poa annua* etc.

Early Seed Setting

Like other crop plants, weeds are also having the tendency to reproduce. Weeds have rapid seedling growth and ability to reproduce when young e.g. Redroot Pigweed can flower and reproduce when it is less than eight inches tall. Weeds have quick maturation period or takes only a short time in the vegetative phase e.g. **Canada thistle** can produce mature seeds in two weeks after flowering. e.g. **Russian thistle** seeds can germinate very quickly between 28^0-$110\ ^0$F in late spring.

Repeated Germination in Different Phases

Weeds appear in crops without being sown or cultivated. Moreover, these appear in different flushes in the same crop and do not loose association till the harvest of the crop. The appearance of *Trianthema portulacastrum* and *Cyperus rotundus* in kharif (summer) crops, *Phalaris minor* and *Chenopodium album* in wheat and other rabi season crops, in different flushes helps in the persistence of these weeds till the harvest of these crops and even the later flushes escape removal even with the adopted weed control measures and compete with the crop and shed' bulk quantity of seeds for the next year.

Deep Root System

Perennial weeds have very deep root system which is usually below the plough layer and thus, it is very difficult to remove /control all vegetative parts of these perennial species through mechanical means. Roots of *Convolvulus arvensis* may penetrate up to 20 feet deep in soil whereby vegetative parts of *Sorghum halepense, Cyperus rotundus and Cirsium arvense* may penetrate 5-7 feet deep in the soil.

Similarity of Seeds

Seeds of *Cichorium intybus* (kashni) are morphologically similar to that of berseem (*Trifolium alexandrium* L.) seeds and these cannot be identified. Similarly seeds of few weeds like *Convolvulus arvensis* (Lehli/hirankhuri),

Avena ludoviciana (wild oats), *Rumex ispinosus* (Kandyali palak) etc. are of the same size as that of wheat and these weed" seeds cannot be separated out from grains of wheat by any method. Hence, use of weed seed contaminated crop seed results in infestation of field with new weed species which were not present during previous years.

Classification of Weeds

Out of 2,50,000 plant species, weeds constitute about 250 species, which are prominent in agricultural and non-agricultural system. Under world conditions about 30000 species is grouped as weeds.

Weeds can be Classified as Under

According to their economic importance:

1. Relative weeds: Unwanted plants which have some economic importance but are called weeds because these are growing out of their place e.g. *Saccharum munja* and *Typha latifolia* are used in cottage industry and few weeds like *Phalaris minor*, *Avena ludoviciana*, *Cynodon dactylon*, *Sorghum halepense*, *Echinochloa crusgalli*, etc. can be used as fodders.
2. Absolute weeds: Unwanted plants which have no economic value and growing out of their proper place are called absolute weeds. i.e. *Euphorbia hirta, Euphorbia microphylla, Amaranthus spinosus, Portulaca oleracea*, etc.

According to Rheir Life Cycle (Ontogeny)

Weeds are also grouped according to their life cycles (seed to seed) into three categories:

1. Annual weeds: Weeds that live only for a season or a year and complete their life cycle in that season or year are called as annual weeds.

These are small herbs with shallow roots and weak stem. Produces seeds in profusion and the mode of propagation is commonly through seeds. After seeding the annuals die away and the seeds germinate and start the next generation in the next season or year following.

According to their season of occurrence, annual weeds can further be classified into two groups :

a) Kharif Annuals (Rainy season & Summer annuals)

These weeds generally appear with the onset of pre-monsoon or monsoon rains i.e. from May and the life cycle is completed earlier or along with the kharif crops i.e. end September upto first fortnight of October. *Trianthema portulacastrum, Eleusine aegyptiacum, Celosia argentea, Digera arvensis, Echinochloa crusgalli* etc. are the common examples.

b) Rabi Annuals (Winter annuals)

These weeds start their growth (germination and emergence) with the fall in temperature i.e. end of October, flower and set seeds before the summer season starts (before May). Their life cycle coincides with the life cycle of rabi crops. The weeds including *Chenopodium album, Medicago denticulata, Trigonella polycerata, Phalaris minor, Avena ludoviciana, Poa annua, Asphodelus tenuzfolius* etc. are the common rabi annuals.

Most Common Field Weeds are Annuals. The examples are

a) Monsoon annual: *Commelina benghalensis* and *Boerhavia erecta*

Fig. 1.1.1
Commelina benghalensis

Fig. 1.1.2
Boerhavia erecta

Fig. 1.1.3
Chenopodium album

b) Winter annual: *Chenopodium album*

2. Biennial weeds : These weeds require at least two seasons to complete their life cycle. In the first season they complete their vegetative growth and store food while in the second season flowering and seed setting takes place. Examples: *Daucus carota, Launea nudicaulis, Cirsium vulgare, Alternanthera echinata* etc.

3. Perennial weeds : These weeds grow for two or more years in the field crops. Depending upon the species, in the first year they generally do not produce seeds but start seed production thereafter. They are capable of propagating by seeds, rhizomes, tubers, stolons, roots, suckers etc. Summer season perennials (*Sorghum halepense, Cyperus rotundus, Cynodon dactylon* etc.) cease their vegetative growth during extreme winters whereas winter perennials (*Cirsium arvense, Convolvulus arvensis*) cease growth during months of high temperature. Perennial are further classified into

i) Simple perennials: Plants propagated only by seeds.eg. *Sonchus arvensis*

ii) Bulbous perennials: Plants which possess a modified stem with scales and reproduce mainly from bulbs and seeds. eg.*Allium* sp.

iii) Corm perennials: Plants that possess a modified shoot and fleshy stem and reproduce through corm and seeds. eg.*Timothy* (*Phleum pratense*)

Fig. 1.1.4 *Sonchus arvensis*

Fig. 1.1.5 *Phleum pratense*

iv) Creeping perennials: Reproduced through seeds as well as with one of the following.

a) Rhizome: Plants having underground stem – *Sorghum halapense*

b) Stolon: Plants having horizontal creeping stem above the ground-*Cynodon dactylon*

Fig. 1.1.6 *Sorghum halapense*

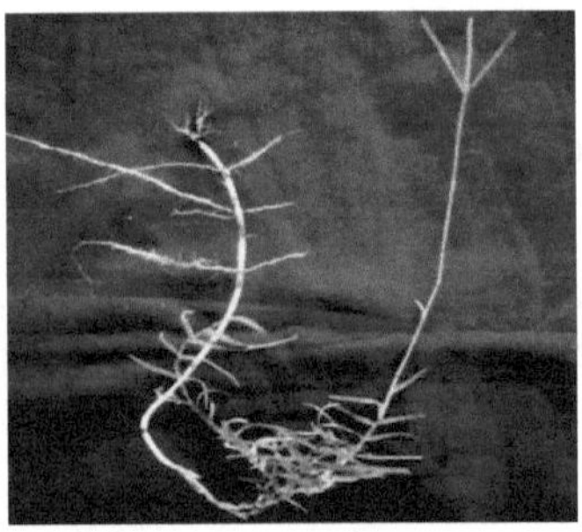

Fig. 1.1.7 *Cynodon dactylon*

c) Roots : Plants having enlarged root system with numerous buds-*Convolvulus arvensis*

d) Tubers: Plants having modified rhizomes adapted for storage of food-*Cyperus rotundus*

Fig. 1.1.8 *Convolvulus arvensis*

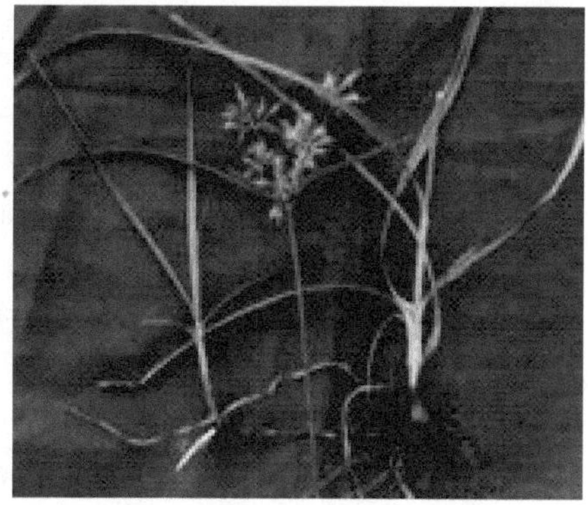

Fig. 1.1.9 *Cyperus rotundus*

According to Their Ecological Affinities

1. Wetland Weeds

They are tender annuals with semi-aquatic habit. They can thrive as well under waterlogged and in partially dry condition. Have lower requirement for light and oxygen than dryland weeds. Anaerobic condition favour most of the wetland weeds. Mostly annual. Most of the weeds in wetland are heavy seed producers. Ability to float in water is achieved by hollowness of the stem. Propagation is chiefly by seed. e.g. *Ammania baccifera* and *Eclipta alba.*

Fig. 1.1.10 *Ammania baccifera*

Fig. 1.1.11 *Eclipta alba*

The following table gives the important weeds of wetland condition.

S.No.	Scientific Name	Family	Tamil Name	English Name
1.	*Ammania bacifera*	Lythraceae	Neermel neruppu	Red stem
2.	*Asteracantha longifolia*	Acanthaceae	Neer mulli	Creeping spider ling
3.	*Centella asiatica*	Umbelliferae	Vallarai	Indian penny wort
4.	*Cyperus diflormis*	Cyperaceae	Vattakorai	-
5.	*Cyperus irria*	Cyperaceae	Oosi Korai	Yellow nut sedge
6.	*Eclipta alba*	Compositae	Karisalanganni	False daisy
7.	*Echinochloa colona*	Poaceae	Kudiraivalipul	Jungle rice/world rice weed
8.	*Echinochloa Crusgalli*	Poaceae	-	Banyard grass
9.	*Fimbristylis miliaceae*	Cyperaceae	Poong Korai	Hoorah grass
10.	*Eichhomia Crassies*	Pontederaceae	Akasa thamarai	Water Iyacinth
11.	*Imperata Cylindrica*	Poaceae	-	-
12.	*Lippia nodiflora*	Verbanaceae	Poduthalai.	-
13.	*Ludwigia parviflora*	Onagraceae	Neer Kirambu	-
14.	*Marsilia quadrifoliata*	Marsileaceae	Arakeerai	Papper work
15.	*Mimosa pudica*	Mimosidae	Thotta churungi	The sensitive plant
16.	*Monochoria vaginalis*	Pontederiaceae	Neer thamarai	-
17..	*Moniera cuneifolia*	Scrophulariaceae	Neer piranni	Thyme-leaved gratiala
18.	*Phyllanthus niruri*	Euphorbiaceae	Keelanelli	Gripe weed
19.	*Panicum repens*	Poaceae	Inji pullu	Torpedo grass
20.	*Rotala densiflora*	Lythraceae	-	-
21.	*Sphaeranthus indicus*	Compositae	Kotta Karanthai	-
22.	*Brachiaria* Spp.	Poaceae	-	Single grass

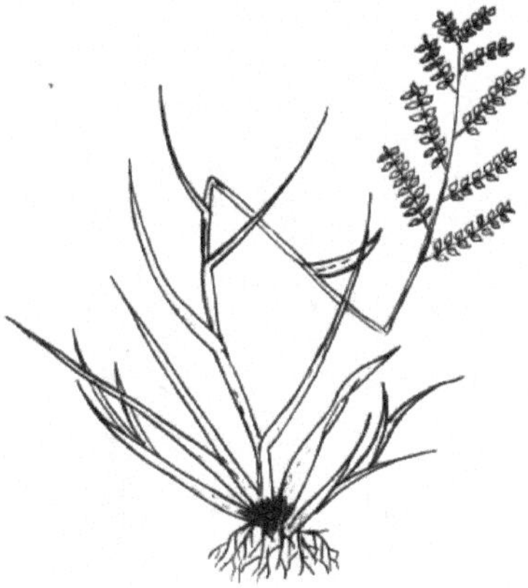

Scientific Name : *Echinochloa crusgalli*
Family : Poaceae
Tamil Name : Uarsnampul
English Name : Barnyard grass

Scientific Name : *Echinochloa colona*
Family : Poaceae
Tamil Name : Kudiraivalipul
English Name : World rice weed

Scientific Name : *Cyperus rotandus*
Tamil Name : Korai Khizhangu
Family : Cyperaceae
Habitat : Cultivated fields and wastelands

Scientific Name : *Fimbristylis miliaceae*
Tamil Name : Poong Korai
Family : Cyperaceae
Habitat : Swampy tanks and rice fields

Scientific Name	:	*Ludwigia parviflora*
Tamil Name	:	Neer Kirambu
Family	:	Onagraceae
Habitat	:	Rice fields

Scientific Name	:	*Lippia nodiflora*
Family	:	Verbanaceae
Habitat	:	Field bunds

Common Name	:	Bermuda grass
Scientific Name	:	*Cynodon dactylon*
Family	:	Poaceae
Habitat	:	Cultivated fields and waste lands

Scientific Name	:	*Alternanthera echinata*
Tamil name	:	Mull Ponnangany
Family	:	Amarantaceae
Habitat	:	Wetland

Scientific Name	:	*Michoc* Neerpirami.
Tamil Name	:	*Moniera cuneifolia*
Family	:	Scrophulariaceae
Habitat	:	Wet places

Common Name	:	Rotala
Scientific Name	:	*Rotala densiflora*
Family	:	Lythraceae
Habitat	:	Moist places

Scientific Name	:	*Ammania bacifera*
Tamil Name	:	Neermael Neruppu
Family	:	Lythraceae
Habitat	:	Wet places

Scientific Name	:	*Eclipta alba*
Tamil Name	:	Karisalanganni
Family	:	Compositae
Habitat	:	Moist places and roadsides

Scientific Name : *Asteracantha longifolia*
Tamil Name : Neer mulli
Family : Acanthaceae
Habitat : Moist places

Scientific Name : *Centella asiatica*
Tamil Name : Vallarai Keerai
Family : Umbelliferae
Habitat : Moist situations

2. Garden land weeds (Irrigated lands)

These weeds neither require large quantities of water like wetland weeds nor can they successfully withstand extreme drought as dryland weeds e.g. *Trianthema portulacastrum* and *Digera arvensis*

Fig. 1.1.12
Trianthema portulacastrum

Fig. 1.1.13
Digera arvensis

List of some commonly grown gardenland weeds

S.No.	Scientific Name	Family	Tamil Name	English Name
1.	*Acalypha indica*	Euphorbiaceae	Kuppai meni	Copper leaf
2.	*Amaranthus viridis*	Amaranthaceae	Kuppaikeerai	Pig weeds/slender pig weed
3.	*Boerhaevia diffusa*	Nyctaginaceae	Mukkuratai	Spreading hog weed
4.	*Celosia argentea*	Amaranthaceae	Pannai keerai	white cock's comb (Silver spiked Cock's comb)
5.	*Chloris barbata*	Poaceae	Mayil kondai pul	Finger grass
6.	*Cleome viscosa*	Capparidaceae	Nai kadugu	wild mustard
7.	*Commelina benghalensis*	Commelinaceae	Kannan Keerai	Kanchara, kana, water grass
8.	*Cyperus bulbosus*	Cyperaceae	Silandi	Bulb grass
9.	*Cyperus rotundus*	"	Korai kilangu	purple nut sedge
10.	*Cynodon dactylon*	Poaceae	Arugu	Devil's grass/ Bermuda grass
11.	*Dactyloctenium aegyptium*	Poaceae	Kakakal pul	Crow foot grass
12.	*Digitaria Sanguinalis*	"	Arisipul	Crab grass
13.	*Digera arvensis*	Amaranthaceae	Thoyya keerai	Amaranthus
14.	*Gynandropsis penta phylla*	Capparidaceae	Kattukadugu	Carvella seed
15.	*Datura fastuosa*	Solanaceae	Umatham	Thorn apple
16.	*Gomphrena decumbens*	Amaranthaceae	Neervadamalli	
17.	*Hibiscus micranthus*	Malvaceae	Nithyamalli	-
18.	*Hibiscus vitifolicus*	Malvaceae	Manja tutter	-
19.	*Orabanche cernua*	Oranbanchaceae	Puhaiyilai Kalan	-Broom rape
20.	*Oxalis corniculata*	Gerniaceae	Puliyarai Keerai	The Indian sorrel
21.	*Portulaca oleracea*	Portulacaceae	Paruppu Keerai	Indian purslane
22.	*Portulaca quadrifida*	"	Pasalai Keerai	-
23.	*Rhyncosia minima*	Leguminosae	Chitt - avarai	-
24.	*Solanum nigrum*	Solanaceae	Manathakkali	-Garden night shade
25.	*Spergula arvensis*	Caryophyllaceae	Thadi Keerai	Spurry
26.	*Striga lutea*	Scrophularaceae	Pallipoondu	witch weed
27.	*Trianthema portula-castrum*	Aizoaceae	Saranai	
		Punchure vine/		
	Horse purselane			
28.	*Trianthema triquetia*	Aizoaceae	Siru Saranai	-
29.	*Trianthema decandia*	Aizoaceae	Vellai Sarani	-
30.	*Vicoa indica*	Compositae	Mookuthi Poodu	-
31.	*Vernonia Cinrea*	Compositae	Mookuthi poondu	Ash coloured flea bane
32.	*Clitoria ternata*	Leguminaceae	Kuvalai	Conch flower
33.	*Ageratum conyzoides*	Compositae	-	Goat weed

Scientific Name : *Euphorbia hirta*
Tamil Name : Amman Pacharisi
Family : Euphorbiaceae
Habitat : Cultivated fields and waste lands

Scientific Name : *Amaranthus viridis*
Tamil Name : Kuppai Keerai
Family : Amaranthaceae
Habitat : Cultivated fields and waste lands

Scientific Name : *Celosia argentea* Linn.
Tamil Name : Pannai Keerai
Family : Amarantaceae
Habitat : Sandy or irrigated clay loam

Scientific Name : *Digeria arvensis*
Tamil Name : Thoyya keerai
Family : Amaranthaceae
Habitat : Cultivated fields and waste lands

Scientific Name : *Tridax procumbens*
Tamil Name : Vettuka Poondu
Family : Compositae
Habitat : Waste places and along edges of cultivated fields

Scientific Name	:	*Orabanche cernua*
Tamil Name	:	Pokayilai - Kalan
Family	:	Orobanchaceae
Habitat	:	Cultivated places

Scientific Name	:	*Cleome viscosa*
Tamil Name	:	Manja Kadugu / Nai Kadugu
Family	:	Capparidaceae
Habitat	:	Waste places and cultivated fields

Scientific Name	:	*Trianthema portulacastrum*
Tamil Name	:	Saranai
Family	:	Aizoaceae
Habitat	:	Wastelands and cultivated places in the plains

Scienitifc Name : *Corchorus* sp.
Tamil Name : Perum Punnakku Poondu
Family : Umbelliferae
Habitat : Irrigated areas

Scientific Name : *Solanum nigrum*
Tamil Name : Manathakkali
Family : Solanaceae
Habitat : Garden night shade

Scientific Name : *Commelina benghalensis*
Family : Commelinaceae
Tamil Name : Kannan Keerai

Scientific Name : *Ageratcim conyzsides*
Family : Compositae
Tamil Name : Goat weed

Scientific Name : *Portulaca olerocea*
Family : Portulacaceae
Tamil Name : Paruppu Keerai

Scientific Name : *Digitaria songuinalis*
Family : Poaceae
Tamil Name : Arisipul
English Name : Crab grass

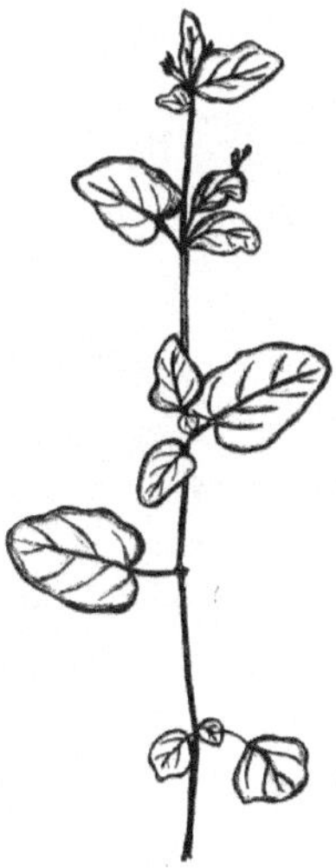

Scientific Name : *Boerhavia diffusa*
Family : Nyctaginaceae
Tamil Name : Mukkuratai

3. Dryland Weeds

These are usually hardy plants with deep root system. They are adapted to withstand drought on account of mucilaginous nature of the stem and hairiness. In the drylands of Tamil Nadu, Sorghum, groundnut, pulses and cotton are the dominant crops. The detailed survey of the dominant weed flora on farmer's field in dryland areas showed that weeds belonging to 'Poaceae' family are dominant in crops like millets and maize, while broad leaved weeds are most abundant in pulse and vegetable.

Adaptive Mechanisms of Dryland Weeds

1. Deep root system for effective moisture extraction.
2. Long dormancy period
3. Efficient seed dispersal mechanism
4. Modification of above and below ground parts, presence of cuticle, vascular bundles, waxy layers, etc. e.g. *Tribulus terrestris* and *Argemone mexicana*.

Fig. 1.1.14 *Tribulus terrestris*

Fig. 1.1.15 *Argemone mexicana*

List of major weeds that are grown in dryland areas are as follows :

S.No.	Scientific Name	Family	Tamil Name	English Name
1.	*Eleusine indica*	Poaceae	—	Goose grass
2.	*Anagallis arvensis*	-Primulaceae	—	Scarlet pimpernal
3.	*Aristacochia bracteata*	Aristolochiaceae	Aduthinna Palai	—
4.	*Corchorus trilocularis*	Tilliaceae Vazhukkai poondu	Punnakku or	Jew's Mallow
5.	*Eragrostic tenui folia*	Poaceae	—	Lovegrass
6.	*Euphorphia hirta*	Euphorbiaceae spurge	Amman pacharisi	Spurge/Garden
7.	*Tridax procumbans*	Compositae	Vettukaya poondu	Tridex
8.	*Solanum elaegnifolium*	Solanaceae Kathiril	Kattu Kandan	White horse nettle
9.	*Solanum xanthocarpum*	Solanaceae	Kanda Kathiri	Night shade
10.	*Oldenlandia umbellata*	Rubiaceae	Saya veir madder	Chayroot; Indian
11.	*Sida spinosa*	Malvaceae	—	Spiny sida
12.	*Physallis minima*	Solanaceae	-Thakkali	Ground cherry
13.	*Osimum canum*	Labiatae	Naithulasi	-The hoary basi
14.	*Leucus utricae folia*	Labiatae	Perum thumbai	—
15.	*Leucus aspera*	,,	Thumbai	Guraa

Scientific Name : *Trichoderma indicum*
Tamil Name : Kazhuthai Thumbai
Family : Boraginaceae
Habitat : Waste places

Scientific Name : *Convolvulus arvensis*
Tamil Name : Bhoomi Chakara Poondu
Family : Convolvulaceae
Habitat : Cultivated and waste places

Scientific Name : *Gomphrena decumbens*
Tamil Name : Neer-Vada Malli
Family : Amarantaceae
Habitat : Roadsides

Scientific Name : *Striga lutea* Lour
Tamil Name : Palli Poondu
Family : Scrophulariaceae
Habitat : Dry grassy places

Scientific Name : *Merremia emarginata*
Tamil Name : Elikathu Keerai
Family : Convolvulaceae
Habitat : Black cotton soils & also in damp places

Scientific Name : *Physalis minima*
Tamil Name : Thakkali
Family : Solanceae
Habitat : Waste places and Cultivated fields

Scientific Name : *Oldenlandia umbellata*
Tamil Name : Saya Vair
Family : Rubiaceae
Habitat : Dry, sandy and grassy places

Scientific Name : *Sida spinosa*
Tamil Name : Spiny Sida
Family : Malvaceae
Habitat : Wasteland and dry channels

According to the Soil Type (Edaphic)

1. Weeds of black cotton soil: These are often closely allied to those that grow in dry condition. eg. *Aristolochia bracteata*
2. Weeds of red soils: They are like the weeds of garden lands consisting of various classes of plants.eg.*Commelina benghalensis*
3. Weeds of light, sandy or loamy soils: Weeds that occur in soils having good drainage. eg.*Leucas aspera*

Fig. 1.1.16. *Aristolochia bracteata*

Fig. 1.1.17. *Leucas aspera*

Fig. 1.1.18. *Lantana camara*

Fig. 1.1.19 *Spergula arvensis*

According to Their Leaf Morphology

1. Grasses

All the weeds come under the family Poaceae are called as grasses which are characteristically having long narrow spiny leaves with simple, long and tapering, opposite having leaf sheath, leaf blade and venation is parallel. Stem is having nodes and internodes. Root system is generally fibrous and shallow. The examples are *Cynodon dactylon, Avena ludoviciana, Lolium rigidum, Phalaris minor, Poa annua, Echinochloa crusgalli, Eleusine aegyptiacum, Acrachne racemose, Digitaria sanguinalis, Sorghum halepense,* etc.

2. Sedges

The weeds belonging to the family Cyperaceae come under this group. The leaves are mostly from the base having modified stem with or without tubers. The examples are *Cyperus rotundus* and *Fimbrystylis miliaceae.*

3. Broad Leaf Weeds

This is the major group of weeds as all other family weeds come under this except that is discussed earlier. All dicotyledon weeds are broad leaved weeds. Leaves are compound, oblong/ovate also having trifoliate leaves. Stem is well branched having primary, secondary and tertiary branches, flowers may be at the apex or axillary, root system deep and mostly having tap root system. The examples are *Flavaria australacica, Digera arvensis, Tridax procumbens, Cirsium arvense, Chenopodium album, Melilotus alba, Medicago denticulata, Digera arvensis, Celosia argentea, Cleome viscosa, Chorchorus tridens, Trianthema portulacastrum, Cirsium arvense, Tribulus terrestris*, etc.

(Grass)
Fig. 1.1.20 *Cynodon dactylon*

(Sedge)
Fig. 1.1.21 *Cyperus rotundus*

(Broad leaf weed)
Fig. 1.1.22 *Tridax procumbens*

According to Their Botanical Families

1. Gramineae: *Avena ludoviciana, Phalaris minor, Poa annua, Lolium temulentum, Eleusine aegyptiacum, Eleusine indica, Acrachne racemose, Digitaria sanguinalis, Cenchrus catharticus, Sorghum halepense, Saccharum spontaneum, Saccharum munja* etc.
2. Solanaceae: *Solanum nigrum, Solanum xanthocarpum*
3. Amaranthaceae: *Amaranthus viridus, Amaranthus spinosus*
4. Euphorbiaceae: *Euphorbia hirta, Euphorbia microphylla, Celosia argentea*
5. Leguminoseae: *Melilotus alba, Medicago denticulata, Trigonella polycerata, Lathyrus* aphaca
6. Tiliaceae : *Corchorus capsularis*
7. Liliaceae: *Asphodelus tenuifolius*
8. Compositeae: *Carthamus oxyacantha. Cichorium intybus, Cirsium arvense*
9. Convolvulaceae: *Convolvulus arvensis, lpomoea spps, Cuscuta* spp.
10. Chenopodeaceae: *Chenopodium album, Chenopodium murale*
11. Cyperaceae: *Cyperus rotundus, Cyperus bulbulus, Cyperus diformis, Cyperus iria*
12. Cruciferacea: *Sisymbrium irio*

According to Their Place of Occurrence (Habitat)

1. Crop land weeds: All Rabi (winter) and Kharif (summer) weeds. The majority of weeds infests the cultivated lands and cause hindrance to the farmers. eg. *Phalaris minor* in wheat.
2. Weeds of lawns and playgrounds: These mostly grow in undisturbed areas eg. *lmperata* cylindrica, Dicanthium annulatum, Medicago denticulata, Euphorbia microphylla, *Cynodon dactylon, Cyperus rotundus.*
3. Weeds of road sides: They are usually hardy, prostrate perennials, capable of withstanding any amount of trampling. E.g. *Alternanthera echinata,Tribulus terestris, Parthenium hysterophorus, Cannabis sativa, Saccharum spontaneum, Saccharum munja, Zizyphus rotundifolia.*

4. Non crop land weeds (also called as industrial weeds): Corners of fields, margins of channels etc., where weeds grow in profusion. eg. *Gynandropsis pentaphylla, Calotropis gigantea, Partheniurn hysterophorus, Argemone mexicana, Solanum xanthocarpum, Ghenopodium murale, Calotropis procera, Crotolaria medicagnea, Zizyphus rotundifolia* etc.
5. Weeds of water channels: *Euphorbia hirta, Xanthium strumarium, Saccharum* spontaneum, Cannabis sativa, Erigeron linifolius
6. Grass land, pasture or rangeland weeds: Weeds found in pasture/grazing grounds. E.g. Indigofera *Enneaphylla, Cynodon dactylon, Cyperus rotundus, Digitaria sanguinalis, Dicanthium annulatum, Medicago denticulate*
7. Aquatic weeds: Weeds which grow in water. *Hydrilla verticillata, Potamegeton* spp.

Fig. 1.1.23 *Gynandropsis pentaphylla*

Fig. 1.1.24 *Calotropis gigantea*

Fig. 1.1.25 *Phalaris minor*

Fig. 1.1.26 *Indigofera enneaphylla*

According to Their Origin

1. Indigenous weeds: All the native weeds of the country are coming under this group and most of the weeds are indigenous. e.g. *Acalypha indica* and *Abutilon indicum.*

Fig. 1.1.27 *Acalypha indica*

Fig. 1.1.28 *Abutilon indicum*

2. Introduced or Exotic weeds: These are the weeds introduced from other countries. These weeds are normally troublesome and control becomes difficult. eg. *Parthenium hysterophorus, Phalaris minor* and *Acanthospermum hispidum*

Fig. 1.1.29 *Parthenium hysterophorus*

Fig. 1.1.30 *Acanthospermum hispidum*

Obnoxious Weeds

Invasive weeds are an important problem for natural and agronomic systems and a major threat to global biodiversity (Wilcove DS *et al.*, 1998). According to the evolution of increased competitive ability (EICA) hypothesis, plants in invasive range allocate more to growth than to defense (Blossey and Notzold, 1995), and consequently the invasive plants perform better than plants of the same species from the native range.Obnoxious weeds, such as *Lantana, Parthenium, Ageratum, Chromolaena, Mikania,* and *Mimosa,* have invaded vast areas of forest, grasslands, wastelands, orchards, and plantation crops across the world. *Parthenium,* one of the seven most difficult weeds of the world, was previously a problem on roadsides and non-cultivated areas in India and is now entering into the field crops.

According to Cotyledon of Weeds

Based on number of cotyledons it possess it can be classified as dicots and monocots.

1. Monocots: All grasses and sedges, which normally produce simple, streight and narrow leaf with veins that are arranged in streight line. eg. *Panicum flavidum, Echinochloa colona.*
2. Dicots: All broad leaf weeds, which produce normally the broad leaf having irregular arrangement of veins. eg. *Crotalaria verucosa, Indigofera viscosa.*

(Monocot weed)

Fig. 1.1.31: *Echinochloa colona*

(Dicot weed)

Fig. 1.1.32: *Crotalaria verucosa*

According to Soil pH

Based on pH of the soil the weeds can be classified into three categories :

1. Acidophile – Acid soil weeds eg. *Rumex acetosella*
2. Basophile – Saline & alkaline soil weeds eg. *Taraxacum* sp.
3. Neutrophile – Weeds of neutral soils eg. *Acalypha indica*

Fig. 1.1.33 *Rumex acetosella*

Fig. 1.1.34 *Taraxacum* sp.

According to Their Fependence on Host

Parasitism is defined as the phenomenon of organism living in, on, or with another living organism to complete its life cycle. The major organ of parasitic weed for attachment and penetration of the host tissue is known as the haustorium.

1. Total parasite: Weed plant which totally depend on host and they cannot prepare their own food due to lack of chlorophyll e.g. *Orobanche* spp. It is a parasitic weed of tobacco, cotton, sunflower, tomato, carrot, sugarcane, sarson etc. The seeds remain viable for 20 years in the soil which mostly germinates when roots of host plants exude certain biochemical (allele chemicals).
2. Semi parasite: Weed plant which do not depend on host for entire life but they depend for part of their life cycle on the host e.g. *Cuscuta* sp. This weed has large seeds with sufficient food reserves to allow the radicle to grow extensively while it is seeking a host plant. *Cuscuta reûexa* and *Cuscuta chinensis* are the parasitic weeds of *zizyphus* and *lucerne* crops respectively. Also *Striga* attaches to the roots of host plant soon after germination but does not emerge from the soil for several weeks. During this period it is totally dependent upon the host plant. Once they emerge, *Striga* plants produce chlorophyll and begin to generate their own assimilates, although water and mineral nutrients are drawn from the host plant.
3. Independent Plants which have no dependence on host and these can synthesise their foods themselves e.g. all annual and perennial weeds of crops and other places e.g. *Phalaris minor*, *Trianthema portulacastrum, Cyperus rotundus* etc.

According to Nature of Stem

Depending upon development of bark tissues on their stems and branches, weeds are classified into two categories.

Based on development of bark tissues on their stems and branches, weeds are classified as woody, semi-woody and herbaceous species.

a) **Woody weeds :** These weeds are very hardy in nature and have woody bark with very hard stem. These weeds includes shrubs and under-shrubs and are collectively called brush weeds. E.g. *Lantana camera* (Lantana), *Acacia arabica* (Acacia) and *Prosopis juliflora.* 'mesquite' , *zizyphus rotundifolia "wild plum' Rubus allegheniensis* 'blackberry'. Control measures for these weeds are entirely different from those of arable weeds.

b) **Semi-woody weeds:** e.g.*Croton sparsiflorus*

c) **Herbaceous weeds:** Weeds have green, succulent stems are of most common occurrence around us. e.g. *Amaranthus viridis, Chenopodium album, Phalaris minor, Trianthema portulacastrum.*

(Woody weed)

Fig. 1.1.35
Prosopis juliflora

(Semi-woody weed)

Fig. 1.1.36
Croton sparsiflorus

(Herbaceous weed)

Fig. 1.1.37
Amaranthus viridis

According to Their Associations

According to this classification there are three types of weeds viz., (1) Seasonm-bound weeds, (2) Crop-bound weeds, and (3) Crop–associated weeds.

The season- bound weeds grow on specific season of the year with disregard to the crop species cultivated. These weeds may be either summer annuals or winter annuals. In the case of perennial weeds, the period of their major vegetative growth is taken as their growing season. For example, *Sorghum halepense is* a summer perennial weed whereas C*irsium arvense* is a winter perennial weed.

The crop bound weeds are those species of weeds which usually parasitise the host crops. They depend for their survival upon their plants for nutrition, particularly or fully. *Cuscuta* sp. 'dodder', *Orobanche* spp. 'boom rape' and S*triga* spp. 'witch weed' are the most common crop-bound weeds of the tropics.

The crop- associated weeds, like crop-bound weeds, are also crop specific but for different reasons. They may be associated with certain crops for one or more of the following reasons.

1. Need for specific micro climate: weeds like *Cichorum intybus* 'chicory' and *Coronobus didymus* 'swine cress', require for their best growth shady, cool, and moist habitat which is amply available in crops like lucerne and berseem. Such weeds, therefore are associated with these crops.
2. Mimicry (Mimic Weed): *Oryza sativa* var. fatua 'wild rice' in paddy fields and *Avena fatua* 'wild oat' and *Phalaris minor* 'canary grass' in wheat and other small grain crops survive because of their similarity in morphology with the host crops. So is true of *Loranthus* in the tea gardens. This mechanism is called mimicry. A weed like wild oat tends to grow to

the height of winter grains and adjusts its ripening time to the crop over a wide varietal range. This kind of mimicry is called phenotypic mimicry.

3. Ready contamination of crop seeds: *Allium* spp. 'wild onion' and 'wild garlic' and *P. minor* mature their seeds at the same height and time as the winter grains and thus, they easily contaminate crop seeds at harvest time.

The crop-bound and crop-associated weeds are recognized separately because they fall easy prey to a weed control measure like crop rotation.

According to Their Specificity

Besides the various classes of weeds, a few others deserve special attention due to their specificity. They are, 1. Poisonous weeds, 2. Parasitic weeds and 3. Aquatic weeds.

1. Poisonous weeds: The poisonous weeds cause ailment on livestock resulting in death and cause great loss. These weeds are harvested along with fodder or grass and fed to cattle or while grazing the cattle consume these poisonous plants. Eg. *Datura fastuosa, D. stramonium* and *D. metel* are poisonous to animals and human beings. The berries of *Withania somnifera* and seeds of *Abrus precatorius* are poisonous.

Fig. 1.1.38

Fig. 1.1.39

Fig. 1.1.40

2. Parasitic weeds: The parasite weeds are either total or partial which means, the weeds that depend completely on the host plant are termed as total parasites while the weeds that partially depend on host plant for minerals and capable of preparing its food from the green leaves are called as partial parasites. Those parasites which attack roots are termed as root parasites and those which attack shoot of other plants are called as stem parasites. The typical examples are;

i) Total root parasite-*Orabanche cernua* on Tobacco

ii) Partial root parasite - *Striga lutea* on sugarcane and sorghum

Fig. 1.1. 41 *Orabanche cernua* on Tobacco

Fig. 1.1.42: *Striga lutea* on sugarcane

iii) Total stem parasite - *Cuscuta chinensis* on leucerne and onion

iv) Partial stem parasite - *Cassytha filiformis* on orange trees and *Loranthus longiflorus* on mango and other trees.

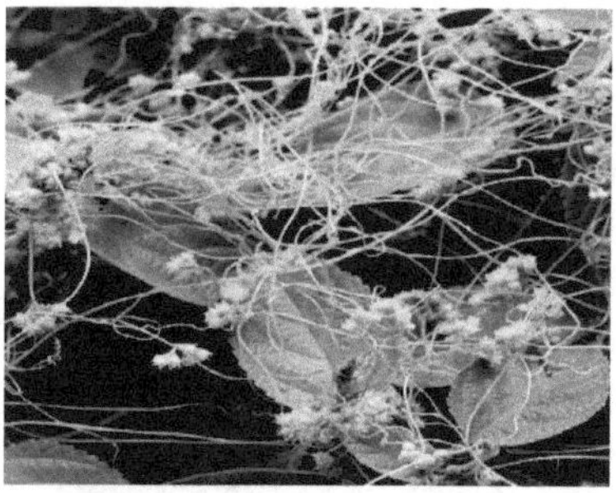

Fig. 1.1.43 *Cuscuta chinensis*

Fig. 1.1.44 *Loranthus longiflorus* on mango

Orabanche spp. (Broom rape; Family: Orobanchaceae)

It is a phanerogamic parasite on tobacco in all tobacco growing tracts in India. It is endemic every year. In the presence of tobacco plants orabanche seeds germinate. *Orobanche* is a parasitic weed, much similar to Striga in its behaviour but with two major differences.

- Its shoot have no under ground life cycle; they emerge from the soil soon after infecting the host roots.
- It lacks green tissue and so it is a total parasite.

The *Orobanche* shoots, called clones are fleshy and cream coloured. At about 3-4 week's age, each shoot produces numerous tiny flowers, viable seeds which are easily dispersed. There are three major parasitic species of broom rape in India. These are *Orobanche cernua* on tobacco and sunflower; *Orobanche ramosa* and *Orobanche aegyptica* on tomato and brinjal and *Orobanche ramosa* in rapeseed and mustard.

Striga spp. (Striga, witch weed; Family: Scrophulariaceae)

Striga is an annual, parasitic herb with a wide range of host crop plants. In India it is most problematic in pearl millet, sorghum, maize and sugarcane. Of the several species of *Striga* in India, *S. asiatica* / *S. lutea* is found most common. Sorghum and pearl millet are parasitised by two different strains of this species. *Striga* produces millions of dust like seeds with high degree of viability. The seeds germinate in response to stimulus from root exudates of the host plants, and the emerging radicles form haustoria on the host roots to establish contact with their vascular bundles to tap food material and water from them. After living underground as a complete parasite for 3-4 week, striga emerges through the soil as a green plant; now depending upon its host only for water and mineral nutrients. At this stage, it grows as semi-parasite. The affected host crop develops symptoms of drought, and its growth is stunted. Application of extra irrigation and nitrogen may tend to revive the crop temporarily.

Species		Host plant
Striga densiflorus	-	Sorghum
Striga lutea	-	Sugarcane
Striga hirsute	-	Pearl millet
Striga hermonthica	-	Maize

Cuscuta spp. (Dodder; Family: Convolvulaceae)

It is a wiry plant and total stem parasite. The wiry stems of *Cuscuta* go around the host plant and slowly cover it completely. The two species are *Cuscuta campestris*, *Cuscuta trifolii*. It develops roots at early stages of development and once attached to the host, the root disintegrates. It is rich in carotenoid pigment which imparts it a golden yellow colour. Stem is thread like, cylindrical and much branched. It forms dense masses which smother host plants. It also develops haustoria masses which penetrate the host plant and extract water and nutrients. In the absence of host plant it dies off. The associated crops are lucerne, lemon, ornamental trees, cowpea, niger, greengram, blackgram, chilli etc.

Contrary to *Striga* and broom rape, the *Cuscuta* seeds do not require any stimulus from its host plant for their germination. Adequate soil moisture and right temperature alone are good enough to germinate them. The *Cuscuta* seedlings emerge from soil and form haustoria on the host plant shoots to share its food and water. The affected host plant dies off slowly due to chiefly water stress. When the cuscuta seedlings fail to find a suitable host plant in viscinity they soon die-off due to starvation.

Loranthus (Bird vine)

It is a semi-stem parasite on trees such as citrus, mango, teak, casuarina etc. The sticky seeds are attractive to birds and are spread by them. This vine attacks rose wood, sandal wood and wattle plantation in Nilgris.

3. Aquatic Weeds

Weed plants growing either in water or having very high requirements for moisture are termed as aquatic weeds. Aquatic plants are also referred to as uunwanted plants, which grow in water and complete at least a part of their life cycle in water are called as aquatic weeds. They are further grouped into four categories as submersed, emersed, marginal and floating weeds.

Types of Aquatic Weeds

The plants that grow in ponds can be categorized into two groups. The algae are primitive plants that have no true roots, stems, or leaves and do not produce flowers or seeds. Algae can be categorized as phytoplankton, filamentous algae, or muskgrass (which resembles a higher aquatic plant).The higher aquatic plants are more advanced and usually have roots, stems, and leaves and produce flowers and seeds. Higher aquatic plants can either be submerged, emergent, or floating.

Classification

Aquatic weeds are often classified according to their habitat as follows:

Common weeds species of aquatic ecosystem.

I. Water weeds (True aquatic weeds)	1. Surface weeds	a. Free floating weeds	Water hyacinth- *Eichhornia crassipes* Water fern - *Salvinia* spp. Water lettuce - *Pistia stratiotes*
		b. Anchored weeds	Water lily - *Nymphea* spp. Lotus-*Nolumbo nucifera*
	2. Submerged / weeds	a. Free floating weeds	Bladder wort, Coon tails
		b. Anchored weeds	Hydrilla - *Hydrilla verticillata Potamogeton* spp.
	3. Emerged weeds	a. Anchored weeds	Cat tails - *Typha latifolia* Arrow heads - *Sagittaria* spp.
	4. Dispersed weeds	-	Unicellular algae - Phytoplankton Multicellular algae - Scum algae
II. Semi-aquatic weeds	1. Shore-line weeds	-	Nut sedge - *Cyperus iria, Marsilea quadrifoliata / M.minuta*
	2. Ditch weeds	-	*Eichhornia crassipes*, *Hydrilla* spp., *Typha* spp., etc.
III. Bank weeds (Ditch-bank weeds)	-	-	Paspalum spp., Cyperus spp., Typha spp., *Scirpus* spp., etc.
IV. Marshland and swamp weeds	-	-	Mangrove trees
V. Phreatophytes	-	-	Lantana, Cotton wood, Willows and Wild rose.

I. Water Weeds (True aquatic weeds)

These weeds grow only in free water and if the water is drained their aerial growth dries away, but the rhizomes or spores survive in the bottom mud. *e.g., Eichhornia crassipes, Hydrilla verticillata, Nolumbo nucifera.*

1. Surface Weeds

a) Free floating weeds: These weeds have leaves that float on the water surface either singly or in cluster. Some weeds are free floating and some rooted at the mud bottom and the leaves rise and fall as the water level increases or decreases. Eg. *Eichhornia crassipes, Pistia stratiotes and Salvinia*

Fig. 1.1.45 : *Eichhornia crassipes*

Fig. 1.1.46 *Salvinia* sp.

b. Anchored Weeds

Nymphaea pubescens.

2. Submersed weeds: These weeds are mostly vascular plants that produce all or most of their vegetative growth beneath the water surface, having true roots, stems and leaves. The greater parts of submerged weeds grow under water and many species possess some out-of-water leaves and inflorescence also. Most of the submerged weed species fail to grow, when water is deeper than 3 m since not enough light reaches to lower depths. *e.g., Hydrilla verticillata, Myriophylum spicatum* (watermol foil) and *Ceratophyallum demassum.*

a) Free floating weeds: Bladder wort, Coon tails

b) Anchored weeds: *e.g., Hydrilla verticillata, and Potamogeton* spp.

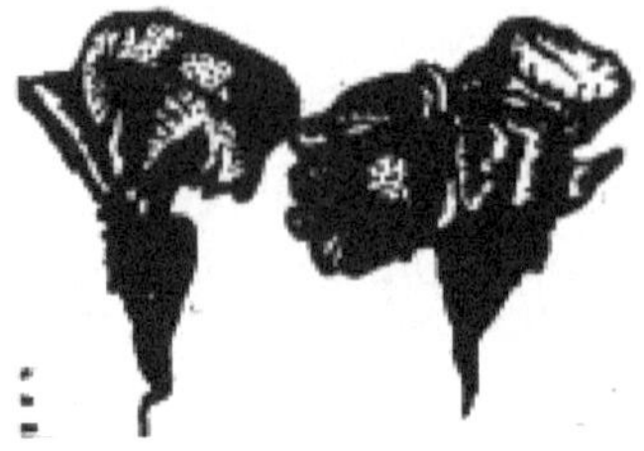

Salvinia

Hydrilla verticillata

Fig. 1.1.47 *Utricularia stellaris*

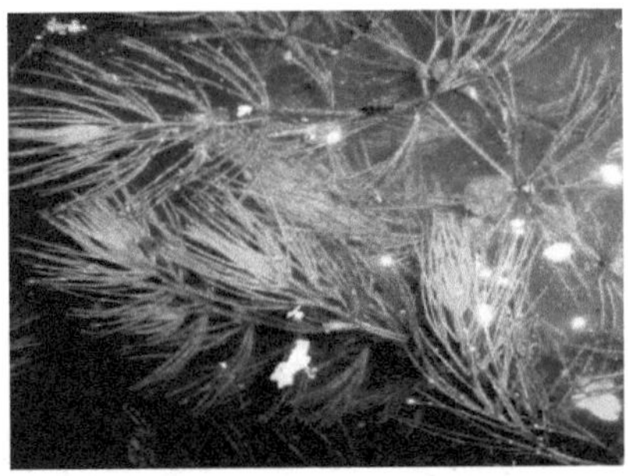

Fig. 1.1.48 *Ceratophyllum demersum*

3. Emerged Weeds

a) Anchored weeds: Cat tails-*Typha latifolia,* Arrow heads-*Sagittaria* spp.

These plants are rooted in the bottom mud, with aerial stems and leaves at or above the water surface. The leaves are broad in many plants and sometimes like grasses. These leaves do not rise and fall with water level as in the case of floating weeds. Eg. *Nelumbium speciosum* and *Jussieua repens.*

Fig. 1.1.49 *Nelumbium speciosum*

Fig. 1.1.50 *Jussieua repens*

Marginal weeds: Most of these plants are emersed weeds that can grow in moist shoreline areas with a depth of 60 to 90 cm water. These weeds vary in size, shape and habitat. The important genera that comes under this group are *Typha, Polygonum* and *Cephalanthus, Scirpus, etc.*

Fig. 1.1.51 *Typha* sp.

Fig. 1.1.52 *Polygonum* sp.

4. **Dispersed weeds:** Unicellular algae-Phytoplankton, Multi cellular algae - Scum algae. The dispersed weeds grow and spread throughout the volume of water. Eg. Algae.

Algae

Phytoplankton

These algae are microscopic simple plants (sometimes referred to as microalgae) suspended in water or forming floating scums of near-microscopic colonies on pond surfaces (Robert, 2014). Communities of phytoplankton are called the "bloom." There are hundreds of species of phytoplankton and identification of the different species is difficult, requiring a microscope (Durborow *et al.*, 2008). Phytoplankton are the most common type of plants found in ponds. Moderate densities of phytoplankton are desirable in ponds because they shade the pond bottom, preventing establishment of more troublesome types of weeds. This competitive relationship also works in the other direction; macrophytes such as alligator weed (*Alternanthera philoxeroides* [Mart.] Griseb) can have algicidal properties that prevent phytoplankton from establishing in a pond. Phytoplankton develop into a weed problem when they become excessively abundant or when certain undesirable species become dominant in the community. Excessive phytoplankton abundance causes serious water quality problems, such as frequent periods of dangerously low concentrations of dissolved oxygen; this low dissolved oxygen occurs when respiration of the heavy bloom exceeds its oxygen production, usually at night and during cloudy weather when photosynthesis is low. This problem is often associated with dense blooms of blue-green algae (Fig as shown below). The blue-green bloom floats to the top of the pond forming a scum that can block sunlight and prevent proper photosynthesis. These blooms can also cause an off-flavor problem in fish raised in aquaculture ponds.

Blue-green algae (cyanobacteria) forming a surface scum

II. **Semi-aquatic Weeds**

1. Shore-line Weeds

Aquatic weeds that inhibit only shore-line water body are called shore-line weeds. These weeds may either germinate from substratum under the water or may start on the shores and slowly creep into shore-line water. *e.g., Scirpus* spp., *Cyperus iria, Potamogeton nodosus, Marsilea quadrifoliata, M.minuta* etc.

2. Ditch Weeds

These weeds invade ditches, both drainage and irrigation channels. The ditch weeds are found in canals, roadside ditches etc., *e.g., Eichhornia crassipes, Hydrilla* spp., *Typha* spp., etc.

III. **Bank Weeds (Ditch-bank weeds)**

These are deeply rooted and sturdy weeds that infect dry embankments and banks of water bodies. Ditch-bank weeds have well developed vascular transport system in them. *e.g., Paspalum* spp., *Cyperus* spp., *Typha* spp., *Scirpus* spp. etc.

IV. **Marshland and Wwamp Weeds: Mangrove Trees**

V. **Phreatophytes: Lantana, Cotton Wood, Willows and Wild Rose.**

These weeds are found along water streams, irrigation and drainage canals, around reservoirs, on dry beds of rivers and in flood plains. These weeds are woody in nature, tall growing and difficult to control. *e.g.,* Lantana, Cotton wood, Willows and Wild rose.

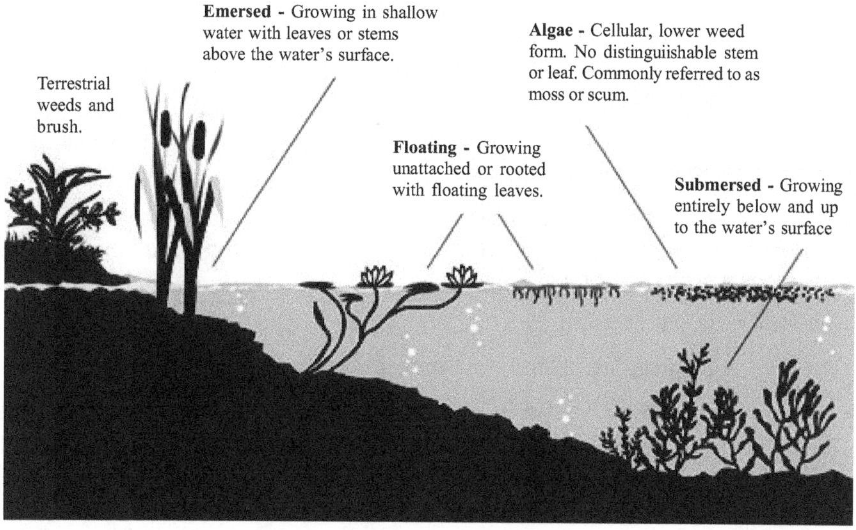

Schematic representation of different types of weeds

Losses Caused by Weeds

Weeds have serious impacts on agricultural production. It is estimated that in general weeds cause 5% loss in agricultural production in most of the developed countries, 10% loss in less developed countries and 25% loss in least developed countries. In India, yield losses due to weeds are more than those from pest and diseases. Yield losses due to weeds vary with the crops. Every crop is exposed to severe competition from weeds. Most of these weeds are self-sown and they provide competition caused by their faster rate of growth in the initial stages of crop growth. In some crops, the yields are reduced by more than 50% due to weed infestation.

Weeds Increase the Cost of Cultivation of Crops

Tillage operations are done to control weeds and it is generally estimated that on an average about 30 per cent of the total expenditure for crop production is on tillage operations and more labour is employed for weeding. This results in increasing cost of cultivation and reducing the margin of net profit.

Poor Quality Produce

Weeds reduce quality of farm, livestock and wool produce.

1. Farm produce When the crop is harvested from a weedy field, the seeds of weeds get mixed with the main crop which results in lowering the quality of the produce .e.g. seeds of weeds in wheat, gram etc. Similarly, bundles of many leafy vegetables like methi or palak contain green plants

of weeds. The contamination of crop produce with various weed seeds not only reduces the quality but the marketable produce generally fetches low price. Market value of wheat is greatly reduced if seeds of *Convolvulus arvensis, Melilotus alba, Chenopodium album, Rumex spinosus, Avena ludoviciana, Phalaris minor* etc. are present in the produce. Similar is the case of rice produce having the seeds of *Echinocloa crusgalli*. If the grains are consumed along with the weed seeds, then sometimes objectionable odour and flavour is imparted to the four. Examples: *Asphodelus tenuifolius* in wheat, wild rice in rice, *Sorghum halepense* in sorghum.

2. Livestock products: Certain weeds such as *Euphorbia simplex, Lepidium sativa, Cleome viscosa* etc. impart an undesirable smell to milk obtained from cattle which graze/feed upon such weeds infested fodder. It is also observed that when milch cattle are fed on fodder infested with such weeds particularly *Lepidium sativa*, a foul smell (odour) not only comes from milk and it also persists in its products also. Some toxic weeds like *Datura stramonium* and *Sorghum halepense* when taken in excess may cause even death of the animals.
3. Wool produce: Certain weeds such as *Cenchrus catharticus, Tribulus terrestris, Achyranthus aspera* and *Xanthium strumarium* may become so much entangled in the hairs/wool of animals (goats/sheep) that it becomes difficult to separate these and thus reduce the quality of the livestock produce (wool) which fetches lower price in the market.

Harbour insect and diseases: Some of the weeds serve as an alternate host for insects and diseases which prove more destructive and harmful for field crops.

i) Bacterial organisms causing bean leaf blight lives on some of the wild legumes.
ii) Organism causing black leg of cabbage thrives on wild mustard (*Sisymbrium irio*).
iii) Downy mildew of lettuce is caused by a fungus that live on weeds of compositae family; wild sunflower and common sow thistle.
iv) Causal organism of loose smut of wheat partly complete its life cycle on *Cynodon dactylon* and *Zizyphus* spps.
v) *Saccharum spontaneum* is the alternate host of fungus causing powdery mildew in maize.
vi) *Cynodon dactylon* and *Cyperus rotundus* are the alternate hosts of grass hopper.
vii) Com borers can live for years in the stems of certain grassy weeds.

viii) Certain virus diseases in lettuce have also been reported from weeds like Russian thistle, nightshade.

Depreciation/Reduce rental value of the land: Certain weeds like *Saccharum spontaneum, Saccharum munja, Zizyphus rotundifolia* and *Cirsium arvense* reduce the sale value of land drastically. Because the cultivation of different crops on such land is only possible after spending heavy expenditure to eliminate these weeds either with machinery or labour.

Weeds check the flow of water: Weeds block drainage and check the flow of water in irrigation channels and field channels thereby increasing the seepage losses as well as losses through overflowing. The irrigation efficiency is also reduced.

Weeds cause quicker wear and tear of farm implements: They get worn out early and cannot work efficiently unless they are properly sharpened or mended.

Impair the health of animals/human beings: Deaths occasionally occur from eating poisonous seeds, berries, tubers or weed plants. Mexican poppy (*Argemone mexicana*) seeds crushed with mustard seeds may lead to death and olindness to people if this contaminated oil is consumed. Milk from cattle feeding on this weed can cause glaucoma in human beings. The *Argemon* toxicity is due to an alkaloid called sanguinarine and oxotriacontanoic acid. Some other poisonous weed seeds are from Rye grass (*Lolium temulentum*), wild mustard (Brassica), Tackla (*Saponaria vaccaria*) and wild garlic (*Allium vineale*). Their seeds may contaminate the food grains and may prove toxic if eaten along with grains.

Carrot grass commonly known as congress grass, gazarghas, chatak chandani, etc. (*Parthenium hysterophorus*) is very harmful for the health of humans and animals. It contains sesquiterpene lactones (Parthenin, hyenim and ambrosin etc.) which induce severe allergic, dermatitis and other symptoms (Swain and Williams, 1977).

In India, 4-7 % population suffers from recognizable clinical symptoms (Tomers and Suba Rao, 1992). When human beings frequently come in contact of this weed i.e. *Parthenium hysterophorus*, it may cause allergy, dermatitis, eczema, black spots and blisters around eyes, burning, rings and blisters over skin, redness of skin and asthma etc. It has been observed that chances of getting sensitized to the weed are high, when a person comes in contact for a period ranging from 3-12 months. In cattle, due to *Parthenium* contact, there is loss of hair, marked depigmentation of skin, development of oedema, ulceration and development of lesions in mouth and intestine.

Certain weeds when fed to cattle along with fodder reduce the acceptability of fodder to the animals such as *Cichorium intybus* in berseem, *Ipomea* spp. and

Cucumus trigonns in bajra or guar. Apart from these weeds, cattle feeding on *Sorghum halepense* during hot months of May and June develop tympani disease due to high concentration of HCN in this weed. However, with onset of monsoon, the plant makes vigorous growth due to which concentration of HCN comes much below the toxic level. Use of *Trianthema* spp. as fodder may cause pre-mature abortion of foetus due to excess nitrate content in this weed.

Health problems caused by weeds to humans

Health problem	Weed
Hay fever and Asthma	Pollen of *Ambrosia* and *Franseria* sp.
Dermatitis	*Parthenium, Ambcosia*
Itching and Inflammation	*Utrica* sp.
African sleeping sickness	Brush weeds
Malaria, encephlitis and filaria caused by mosquito	Aquatic weeds like *Pistia lanceolata, Salvinia auriculata.*

Reduce aesthetic value: Weeds growing along with the lawn grass or other ornamental plants generally give ugly look and reduce the value of the ornamental plants. Infestation of lakes by water hyacinth is another example of nuisance to recreational places.

Allelopathic interaction: There are some biochemical interactions between plants of different species. Severe infestation of some weeds like *Cyperus rotundus, Agropyron repens* and *Trianthema portulacastrum* reduce the germination and growth of many crops because of the production of phytotoxic substances by the roots of these weeds.

Weed secretions are harmful: Heavy growth of certain weeds like quack grass or motha lower the germination and reduce the growth of many crop plants. This is said to be due to the presence of certain phytotoxins in these weeds.

Weeds affect the water life also: Roots of certain water weeds produce toxins/ excess CO_2 which affect the fish life. Even the aquatic weeds utilize oxygen of water in ponds and fish die due to lack of oxygen in water. Decomposition of water weeds/roots may induce the lack of O_2 and fish may die due to deficiency of oxygen or may feel suffocation due to more concentration of CO_2.

Useful Aspects of Weeds

Weeds are not always harmful but may have useful aspects for human beings and animals. Though weeds are hazardous as far as farming is concerned; has beneficial effects too. Many weeds are having economic importance as it can be used for food, fodder, green manure, medicinal and other purposes.

In Scientific Discoveries and Studies

a) Chlorella was used in understanding the details of photosynthesis in higher plants i.e. to know the first products of photosynthesis.

b) *Datura* was used in genetic studies i.e. for understanding the mechanism of inheritance.

c) Kahi (*Saccharum spontaneum*) used in sugarcane breeding.

To check soil erosion: Soil erosion is mainly due to strong wind blowing or water flowings. *Saccharum spontaneum* (kahi) help in reducing wind erosion, which is very common in dry climatic conditions. These situations are ecologically ideal for the growth and development of this weed i.e. *Saccharum spontaneum*. Moreover, *Cynodon dactylon* (khabbal) due to its mat type growing habit helps in minimising water erosion particularly in high rainfall areas. It acts as soil binder and prevents soil erosion. *Panicum repens* is an excellent soil binder; keeps bunds in position and prevents soil erosion.

Medicinal values: Menthol extracted from *Mentha spicata* is used for preparing medicines for human beings. Homeopathic medicine is prepared from *Parthenium* plants in order to cure allergy due to *Parthenium hysterophorus*.

Argemone mexicana- which is used externally for skin diseases. With sugar it is good for stomach troubles. The yellow juice is used in dropsy. Bruised fresh root is good for scorpion sting.

Gynandropsis pentaphylla - Decoction of the seeds and leaves is given in typhoid fever. The oil extracted from seeds is useful in the treatment of leprosy and skin diseases.

Cleome viscosa - They have anthelmintic and carminative properties. The leaves are also used in external applications for wounds and ulcers.

Lonidium spp. - Plant is highly medicinal. It is used in urinary and genital infection. Decoction of the plant improves memory and vitality. It is useful for asthma and leprosy.

Portulaca oleracea - The leaves are a stringent and diuretic. When fresh, they are also used in external cooling applications.

P. quadrifida - Useful in curing low fevers, cough, asthma, eye diseases and skin infections.

Abutilon indicum - The leaves are rich in mucilage and are used as a demulcent tonic.

Malvastrum coranmandelianum - Leaves are applied to inflamed sores and wounds as a healing agent.

Oxalis corniculata - is a good appetiser. Famous as a cure for scurvy.

Biophytum sensitivum - given to children to induce sleep. It is also used for chest complaint and its ash for stomach ache,

Tephrosia purpurea - Decoction of the fruit has vermicidal action. The plant is a laxative and tonic. The root is bitter and given for diarrhea.

Cassia occidentalis - The root is an antidote to poisons and is used in snake bite with pepper.

Mimosa pudica - Whole plant is medicinally used. The root is useful in asthma and diseases of the blood.

Ammannia baccifera - Leaves are used to blister skin. Juice in small doses given to animals in heat to put down sexual appetite.

Trianthema prortulacastrum - is an antidote for alcoholic poison. The plant extract is a remedy for rheumatism.

Centella asiatica -Whole plant is medicinal. It is a tonic, blood purifier and improves the memory. Leaves when used as fodder, increase the secretion of milk in cows.

Oldenlandia umbelleta - The bark in the root contains valuable red dye and is specific for snake bite.

Eclipta alba - The plant is highly used in curing asthma and other diseases. Leaf is a good remedy for scorpion sting. The extract of the plant is useful to remove baldness.

Calotropis gigantea - (i) The root bark and juice have tonic properties. The juice contains 'Mudarine' used as a purgative; (ii) Root is an antidote to poison (iii) Leaves with pepper are used for snake bite.

Solanum xanthocarpum - useful for curing asthma, cough, fever etc.

Ocimum sanctum - infusion of leaves is given in Malaria and in the treatment of skin diseases. The plant is used as mosquito repellent. Continuous daily use will be helplful in family planning.

Achyranthes aspera - The plant is highly medicinal and used for various diseases. The whole plant extract is good for kidney troubles. Antidote for poison of snakes and scorpions.

Aerva lanata - is useful for kidney disorders.

Phyllanthus niruri - Infusion of tender shoots is administered in chronic dysentery and yellow fever (Jaundice).

Food for animals: Some grasses like *Eleusine aegyptiacum, Phalaris minor, Cynodon dactylon, Sorghum halepense* serve as palatable fodder plants. Some broad leaf weeds such as *Trianthema portulacastrum* L. (Itsit/Chupati)

Amaranthus viridis (Chulai), *Digera arevensis* (Tandla) are also used as green fodder. Also seeds of few weeds especially Phalaris minor are used as animal feed.

As pot herbs: Few weed plants such as *Digera arvensis, Amaranthus viridis, Chenopodium album* etc. are used as pot herb as pure or after mixing with mustard plants.

Improve soil fertility: Blue green algae fix atmospheric nitrogen. *Anabeana* and *Nostoc* live in association with Azolla (a fern) and fix atmospheric nitrogen. Some leguminous weeds like *Melilotus alba*,

Medicago denticulata and *Trigonella polycerata* etc. improve fertility level of soil not only by fixing N but due to the addition of organic matter also.

In cottage industry: Leaves of *Typha latifolia* (Dib) are used for preparing mats as well as roaps which are used for tying wheat bundles.

For compost: Water hyacinth (*Eichhornia crassipes*)-an aquatic weed is present in abundance in ponds, lakes, drainage ditches and other water bodies. This weed is very rich in nutrients particularly nitrogen. Compost prepared from this weed is of high value. This weed can also be used as a mulch in widely spaced crops.

For fibre: Few weeds like *Sida cardifolia, Abutilon indicum, Corchorus capsularis* etc. are used for this purpose though the fibre obtained from these weeds is very little and of poor quality.

The economic importance of some prominent weeds are detailed below: *Cyperus rotundus* - Essential oil obtained from the tubers are used in perfumery. The shoots and tubers +are used in medicine as stimulant and vermifuge. It stimulates increased milk secretion if applied externally on breast.

The tubers are edible, with a slightly sweet, nutty flavour, compared to the more bitter-tasting tuber of the related *Cyperus rotundus* (purple nutsedge).

Green manure: The weed *Parthenium hysterophorus* may be used as a substitute green manure. It is better than *Glyricidia maculate* in its nitrogen content and also giving good yield when applied as green manure to rice (Sudhakar and Purusothaman, 1984). Some weeds viz., *Crotalaria verrucosa, C. quinquifolia, Indigofera uniflora, Phaseolus trilobus, Calotropis gigantea, Croton sparsiûorus* are used as green manure.

Food for human beings: Besides the above beneficial weeds, the following weeds are taken as food by human beings either a pan or the whole plant. They are *Amaranthus viridis, Alternanthera triandra, Coccinia indica, Mollugo lotoides, Leptadenia reticulata, Digera.*

1.2

Weed Biology

Weed Biology

Weed biology is the study of the establishment, growth, reproduction, and life cycles of weed species and weed societies/vegetation. Weed biology is an integrated science with the aim of minimizing the negative effects, as well as using and developing the positive effects, of weeds.

Life Cycle - Based on life cycle weeds are classified as annuals, biennials or perennials.

Annuals

Annuals complete their life cycle from seed in less than one year. There are two types: summer and winter annuals. Summer annuals germinate in the spring, mature, produce flowers and seeds and die before fall. Winter annuals germinate in the fall, overwinter in a seedling or rosette stage, mature, produce flowers and seeds, and die in the spring or early summer. Because of the seedling stage, annual weeds are generally easy to control. There are usually few annual weeds present in low bush blueberry fields.

Biennials

Biennials generally complete their life cycle over two years. The first year the seeds germinate and form a basal cluster of leaves and a tap root. The plant overwinters in this stage. During the second year the weed produces a flower stalk, sets seed and dies. Examples of biennial weed are evening primrose and wild carrot.

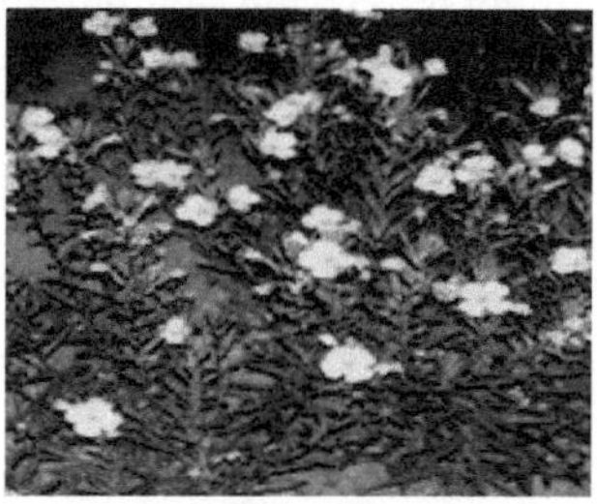

Evening prime rose

Wild carrot

Perennials

Perennial weeds live for more than two years. These weeds are the most common in blueberry fields and generally the most difficult to control. Perennial weeds may reproduce primarily by seed (daisy); by both seed and roots (sheep sorrel); or primarily be vegetative means (bunchberry). Many perennial weeds grow in the same manner as the blueberry plant. Therefore, many of the production practices that promote blueberry growth (e.g. pruning) also promote growth of these weeds.

Growth Characters-Based on growth characters weeds are classified as grasses, broadleaf weeds, ferns and herbaceous or woody weeds.

Broadleaf

Broadleaf weeds are annual, biennial or perennial plants which generally have two leaves (cotyledons) emerging upon germination. The leaves normally have a branching network of veins and the flowers have distinct petals.

Grasses, Sedges and Rushes

Grasses can be annual or perennial plants. They generally have narrow, upright, parallel-veined leaves. Grasses have jointed stems, usually hollow at the internodes and are circular in cross section. Sedges are a large group of perennial (rarely annual) grass-like plants which are common in wet, poorly drained soils. Sedge stems are triangular in cross section, solid, and not jointed.

Rushes are annual or perennial plants similar in appearance to sedges with grass-like tufted leaves common at the plant base. Rush stems are hollow, circular in cross section, and not jointed. Like the sedge, this plant is also common in wet areas or poorly drained soil, but is also found in woodland and open fields.

Ferns

Ferns are primitive perennial plants that do not produce flowers and seeds. Ferns consist of a leaf or frond, a stalk and an expanded blade which may then be further subdivided several times. Ferns spread by long creeping rhizomes and/or by spores.

Fern plant

Fern spores

Propagation of Weeds

Seed is the main and the common source of propagation of plants. The seed production of natural plants (weeds) is much more than economical crop plants. All weed after completion of life cycle have a tendency to shed seeds in the field which become cause of infestation in the future years. Due to high seed production capacity of weeds, their seed reserves goes on increasing in the field. These weed seeds in the conventional farming go as deep as plough layer. The seeds present in the top soil zone will germinate while lying at more depth will not germinate until brought to upper soil layers. Many weed reach 30 cm or more deep, by entering through the soil crevices. *Commelina benghalensis* weed, places its seeds directly inside the soil since it produces both aerial and underground flowers and seeds. Lying within the soil, weed seeds are capable of retaining viability from 2 to 25 years, depending upon the weed species (Corn and Farris, 1983). Some of *Rumex* spp. show 63-90 per cent germination upto 2-7 years with 30 cm deep burial (Hongo, 1994), while seeds of *Echinochloa* spp., *Poa* spp., *Chenopodium* spp. germinated after 25 years of their similar burial in the soil (Suzuki, 1994).

On the other hand, in perennial weeds, the viable seed production capacity is very minimum. Moreover, seed production potential of these perennial weeds is much less than that of others. The propagation of perennials is mostly through underground vegetative parts. However, few perennial weeds like *Convolvulus arvensis, Saccharum spontaneum, Cynodon dactylon, Sorghum halepense* etc. can multiply both sexually and asexually.

Reproduction by Seed

Seed is the life of a plant. All annuals and few perennials multiply through seed. Reproduction by seed is called sexual reproduction.

Through sexual reproduction abundant and small seeds are produced. Annual and biennial weeds depend on seed production, as the sole means of propagation and survival of perennial weeds are less dependent on this mechanism. For example, while *Canada thistle* has been observed to produce as few as 680 seeds per plant, curly dock often produces more than 30,000 seeds per plant. However, all seeds do not germinate at one time due to differences in dormancy, viability, placement depths etc.

Canada thistle

Importance of Seeds

The chief functions of seeds are to perpetuate, multiply and disseminate the species; they are vital to those species lacking vegetative methods of propagation. Many species survive as seeds during unfavourable conditions, such as cold/winter or hot summer remains in dormant condition. In cross pollinated species, genetic viability is maintained by seed production and may produce individuals able to thrive in a variety of habitats. It may also act against the development of herbicide-resistant strains. Most common weed species, however, are self-fertile and one plant may be sufficient to start an infestation. Annual weeds complete their life cycle i.e. seed to seed within one season of a year. Many annual weeds complete their vegetative growth first and then enter into reproductive stage e.g. *Phalaris minor*, *Digera arvensis* etc. However, in few annuals like *Trianthema portulacastrum*, vegetative and reproductive stage goes on side by side. Normally annuals are considered easy to control. However, because of an abundance of dormant seeds and fast growth, annuals are very persistent and they actually cost more to control than perennial weeds.

Time of Flowering

The time duration between germination and the production of first seed depends on the species and on growing conditions. Some annuals e.g. *Trianthema portulacastrum* and *Capsellabur sapestoris* can flower in as little as 6 weeks time from germination in the summer and consequently can persist in frequently cultivated situations, with 2 or even 3 generations in a year. A short life cycle is a very important survival mechanism of a weed in cultivated areas. An additional mechanism is found in *Portulaca oleracea*. The plant is able to continue seed maturation and produce mature seeds even when it is cut after flowering (Gutterman, 1985).

Seed Setting, Development and Maturation

Seed setting in most common weeds is independent of cross-pollination, either because the plants are self-fertile, e.g. *Avena*spp. or because the ovules can develop without fertilisation, e.g. *Taraxacum oûicinale*. Seeds of many weed species, especially of family Compositeae, become viable before they are fully developed. Plants of *Senecio vulgaris* (ground sel) pulled up while in flower can set viable seeds, and if panicles of *Avena* spp. (wild oats) are cut or pulled at the milk stage, seeds are already viable but not yet dormant. *Amaranthus retrofexus*, a facultative short day plant for flowering, was found by Kigel *et al.* (1977 and 1979) to be affected by parental photoperiodic conditions. Seeds matured under short days (8 h) germinated better than seeds matured under long days (16 h).

Number of Seeds Produced by Plants

The persistence of annual and perennial weeds depends mainly on their ability to reinfest the soil. The first infestation of most perennial species depends on seed. Obviously, if we could control the production of seeds, we could eventually eliminate many species. The number of seeds produced by an individual plant varies greatly according to the species. It may be less than 50 in *Veronica hederifolia*, 500 in *Avena fatua* (wild oat), 23400 in *Datura stramonium*, 58363 in *Brassica nigra* and 1,80,220 in *Amaranthus* sp. One plant of *Striga* can produce 40,000 to 60,000 seeds, depending on the species whereas one plant of *Orobanche* can produce about 5.0 lakh seeds.

Weed Seeds in Soil (Seed Bank)

The store of seeds in the soil is composed of those produced in the area (both recent and long term production) and those dispersed from elsewhere.

A weedy field may easily contain 100 millions or more viable weed seed per acre and numbers up to 350 millions per acre have been recorded. In Minnesota, weed seed counts at four different locations on 24 different plots showed between 4.3 million and 133 million seed per acre in the upper six inches (Ashton & Monaco, 1991).

Vegetative Reproduction of Weeds

Vegetative Reproduction

In vegetative (asexual) reproduction, a new plant develops from a vegetative organ such as a stem, root, or leaf. Several modifications of these organs are common in perennial weeds, such as underground stems (rhizomes), above-ground stems (stolons), bulbs, corms, and tubers. Although vegetative structures generally do not survive as long in the soil as do seeds, very small structures can result in a new plant.

Forms of Vegetative Propagation

The most common forms of vegetative propagation found in weed species are:

Stolons and runners: Long slender stems that grow along the soil surface and produce adventitious roots and new shoots. Among several examples are: Bermuda grass (*Cynodon dactylon*), a perennial grass; a large crab grass (*Digitaria sanguinalis*), an annual grass.

Rhizomes : Underground reproductive parts of perennial weeds that produce adventitious roots and shoots, for example, Johnson grass (*Sorghum halepense*) quack grass (*Agropyron repens*), Kans/kahi (*Saccharum spontaneum*).

Tubers: Enlarged terminal portion of rhizomes: These possess extensive storage tissue and axillary buds. Examples are *Cyperus rotundus* and *C. esculentus*, perennial sedges and Jerusalem artichoke (*Helianthus tuberosus*).

Cyperus esculentus– tubers

Bulbs: Underground modified buds consisting of a stem and fleshy scale leaves. Food storage is in the leaves, for example, wild garlic (*Alium vineale*), a perennial herbaceous lily.

Corms: Enlarged, vertical underground stems covered with one or more layers of leaf bases. Food storage is in the stem, for example, bulbous buttercup (*Ranunculus bulbosus*), a perennial herb.

Roots: Many species produce long horizontal roots that give rise to shoots. Examples include canada thistle (*Cirsium arvense*), field bind weed (*Convolvulus arvensis*), perennial herbs and many others.

Stems: Some species produce adventitious roots and new shoots near the tips ofbranches, for example, Rubus sp. (a perennial vine), whereas others produce sprouts from astem base on stump. Examples of the latter are dandelion (*Taraxacum oûicinale*), a perennial herb.

Fragmentation: Spread and establishment of a ramnant by various plant parts, such as excised leaves or stems. This method of propagation may occur with segments of most underground portions of the plant. It is also evident with stems or leaves, for example, purslane (*Portulaca oleracea*), an annual herb.

An important feature of most of the forms of propagation, especially those occurring below ground, is the large number of vegetative propagules that can be produced. Purple nutsedge may reproduce vegetatively from rhizomes, tubers or basal bulbs, though the tuber is the main reproductive and storage organ. Based on field studies, it was estimated that 10 to 15 million tubers hectare^{-1} could be produced within the top 15 cm of soil per year underoptimal conditions.

Dispersal /Dissemination

It refers to the movement of weed seeds/vegetative propagules from one place to another place with the help of different agencies. Annual weeds spread near and far by the dispersal of their seeds while perennial weeds infest distant places through seeds and vegetative parts (bulbs, roots, rhizomes, stolons and tubers etc.). Weeds are dispersed by natural and artificial means because the weed seeds/plant parts have no method of movement. Natural means include wind, water, animals and forceful opening of the fruits/seed pods/capsules. While artificial means are associated with the movement of man and his agricultural materials (crop seeds, grains, feed, hay, straw, manures and machinery) from one place to another place.

Natural Agencies for Dissemination

1. Wind

Some of the weed seeds have many specific adaptations/modifications which help in their dispersal. Some weed seeds being very small, tiny and light in weight (*Calotropis procera, Striga* spp., *Erigeron* spp.) are wind borne and carried to faraway places. The weed seeds which are too heavy to travel may be equipped with parachute like structures, wings and tuftsof hairs (cotton like covering) which improve their aerodynamics and carried from one place to another place. Of these, *Cirsium arvense, Calotropis procera, Sonchus qrvensis* and *Saccharum* spp. are the common examples which are disseminated through wind. Seeds of many weed species such as *Brassica* spp., *Celosia argenrea, Cleome viscosa, Euphorbia hirta* and *Silene conidea* may be rolled for long distances by strong wind because of their lightweight and small size.

Weed seeds and fruits that disseminate through wind possess special organs to keep them afloat. Such organs are :

1. **Pappus -** It is a parachute like modification of persistent calyx into hairs. Eg.Asteraceae family weeds – *Tridax procumbens*

Tridax procumbens

Pappus *seed*

2. **Comose -** Some weed seeds are covered with hairs, partially or fully. Eg. Calotropis sp.

3. **Feathery, persistent styles** - Styles are persistent and feathery. Eg. Anemonesp.

4. **Baloon -** Modified papery calyx that encloses the fruits loosely along with entrapped air. Eg. Physalis *minima*

5. **Wings** - One or more appendages that act as wings. Eg. *Acer macrophyllum*

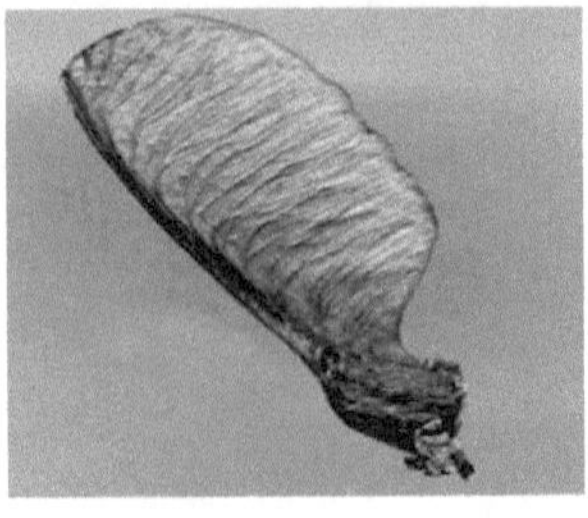

Factors that Influence Wind Dispersal

a) Seed weight
b) Seed shape
c) Structures (wings or pappus)
d) Height of release
e) Wind speed and turbulence

2. Water

Flowing water is capable of transporting almost any type of seeds, which can fioat on surface of water. Depending on the volume and swiftness of water, weed seed may move with surface water, runoff in natural streams, rivers, irrigation and drainage channels. The weed seeds are introduced in water by dropping into the water from mature plants growing along the various flowing water sources. Some of the weed seeds are dropped in the water sources either through wind or through floods or heavy rains. Irrigation water is an important source of scattering weed seeds as most of the weeds grow along sides of irrigation channels due to available moisture and these weeds shed their seeds in water. The most common examples are *Caesulia axillaris, Cyperusiria, Echinochloa crusgalli.*

Weed seed often remains viable in water for several years, creating a "floating seed bank" and allowing weeds to disperse over large areas in moving water. Field bindweed seed, for example, remains over 50 percent viable after being submerged in water for more than 4 years. Some seeds have special adaptations that aid in water travel. The seedpod of Curly dock, for example, is equipped with pontoons that carry the floating seed.

Curly dock

Plant Seed Mature seed

3. Animals

Animals, including man are responsible for scattering weed seeds/vegetative parts. The dispersal units such as barbs (*Achyranthus aspera*), hooks (*Xanthium strumarium*), spines (*Cenchrus catharticus*) and twisted awns (*Avena ludoviciana*), cling to the animal's fur or hair or to man's clothing and taken away to a distant places. Other seeds have cotton like lint or similar structures also cling to the animal fur or man's clothes. Movement of wild animals including rabbits, dogs, jakals, foxes etc. held responsible in dissemination of weed seeds. Even ants carry a huge number of weed seeds. Donkeys eat *Prosophis julifera* pods.

Xanthium strumarium - hook

*Cenchrus*spp - stiff hair

Tribulusterrestris - spines

Birds also carry away weed seeds which stick to their feet or feathers. Few birds particularly sparrows eat on some fruits of weeds (*Lantana camera*) and dispersal of such plants takes place through their droppings. Animals which feed upon weeds growing in pastures or weed infesting fodder / hay / grain feed carry viable seeds from one place to another place since the seed pass through the digestive tract unaffected along their faeces (feces). The weed plants of *Trianthema portulacastrum*, *Celosia argentea, Phalaris minor, Chenopodium album* if fed to the cattle at maturity stage, the seeds of these weeds do not loose viability even after passing through the digestive tracts of animals and the undecomposed dung applied as FYM in the fields carries large quantities of weed seeds.

4. Dispersal by Man

Man disperses numerous weed seeds and fruits with raw agricultural produce. Weeds mature at the same time and height along with crop, due to their similar size and shape as that of crop seed man unknowingly harvest the weeds also, and aids in dispersal of weed seeds. Such weeds are called "Satellite weeds". Eg. *Avena fatua* and *Phalaris minor.*

Avena fatua

Phalaris minor

4. Forceful Opening

Artificial Dissemination

1. Machinery

The dissemination of weed seeds by means of transport system, wagons, trucks, trains, aeroplanes and farm machinery particularly harvesting machines is quite common. Farm machinery can easily carry weed seeds, rhizomes, tubers, corms, stolons and roots from one field another ûeld as these stick/cling to the implements along with the mud/ moist soil etc. Spread of perennial weeds is generally through farm implements as the vegetative parts are carried through mud or soil adhering to the wheels or parts of fann implements. The typical examples are tubers of *Cyperus rotundus*, stolons of *Cynodon dactylon*, rhizomes of *Sorghum halepense* and roots of *Cirsium arvense* which are disseminated to a long distance through farm machinery. Cultivation equipments, tractors and tractor tyres often carry dirt which may include weed seeds. Harvesting combine is a major source of weed seed dispersal. After the harvesting of a crop infested with weeds, the movement of combine to another field without cleaning for weed seeds, take a sizeable quantity of weed seeds to another field. The two classical examples are *Phalaris minor* in wheat and *Echinochloa crusgalli* in paddy which have spread with harvesting combines.

2. Intercontinental Movement of Weeds

Introduction of weeds from one continent to another through is through crop seed, feed stock, packing material and nursery stock. Eg.*Parthenium hysterophorus.*

3. Crop Mimicry Dispersal

Weed seed adaptations to look like crop seed: plant body or seed same size, shape, and morphology as crop. Eg: barnyard grass biotype looking like rice escapes hand weeding and is dispersed with rice, nightshade fruit ("berries") same size, shape as dry beans, harvested and dispersed with beans.

4. As Admixtures with Crop Seed, Animal Feed, Hay, and Straw

Crop seeds, grain feed, hay and straw are also common source of weed seed infestation. If the wheat seed is contaminated with seeds of *Avena ludoviciana, Phalaris minor, Melilotusalba, Chenopodium album* etc. when sown without cleaning may pose a serious problem of these weeds in this crop in the following years. Likewise, the presence of *Echinochloa crusgalli* seeds in paddy seeds may infest the ûeld badly if sown without cleaning. Apart from crop seeds, weeds also spread through grain feed, hay and straw of infested fields.

Persistence of Weeds (Adaptation)

Persistence is an adaptive potential of a weed that enables it to grow in any environment. In an agricultural situation, the cropping system with its (associated habitat) management practices, determines the persistence of weed species. It is largely influenced by climatic, edaphic (soil) and biotic factors, which affect its occurrence, abundance, range and distribution.

Factors Affecting Persistence

A. Climatic Factors

The important climatic factors are light, temperature, rainfall, wind and humidity.

Light: Light intensity, quality and duration are important in influencing the germination, growth, reproduction and distribution of weeds. Photoperiod governs flowering time, seed setting and maturation and on the evolution of various ecotypes within a weed species. Tolerance to shading is a major adaptation that enables weeds to persist.

Temperature: Temperature of atmosphere and soil affects the latitudinal and longitudinal distribution of weeds. Soil temperature affects seed germination and dormancy, which is a major survival mechanism of weeds.

Rainfall: Rainfall has a significant effect on weed persistence and distribution. More rainfall or less rainfall determines reproduction & survival.

Wind: Wind is a principal factor in the dissemination of weeds.

B. Soil Factors

Soil factors are soil water, aeration, temperature, pH and fertility level and cropping system. Some weed species are characteristically alkali plants, known as basophilic (pH 8.5) which can grow well in alkali soils and those grow in acidic soil is known as Acidophiles.

Basophiles	Acidophiles	Neutrophiles
Alkaligrass –*Puccinalia* spp.	*Cynodon dactylon*	common weed
Quack grass-*Agrophyron repens*	*Digitaria sanguinalis*	

Several weed species of compositae family grow well in saline soils. A shift in soil pH, towards acid side due to continuous use of Ammonium sulphate as 'N' source could cause a shift in the weed spectrum.

Many weeds can grow well in soils of low fertility level however, can adapt well to soils of high fertility also. Weeds also has adaptation to moist soil, drought condition etc.

C. Biotic Factors

In a cropping situation, the major effects on weeds are those exerted by the crop as it competes for available resources. Once, a particular weed species is introduced, its persistence is determined by the degree of competition offered by the crop and also the agricultural practices associated with the growing of a crop may encourage or discourage specific weeds.

Eg. Ponding of water -*Cynodon* dies

Repeated cultivation - discourage nut sedge.

Crops that serve as hosts to parasitic weeds, (Sorghum - Striga) crop-induced stimulants are examples of other biotic factors.

Perennation

The process of living through a number of years as a perennial plant. Perennation is also defined as the formation of resting stages to survive unfavorable environmental conditions (e.g. overwintering or other environmental changes that cause a plant to enter dormancy). The second meaning is more close to question.

i) Potato tuber, sweet potato tubers are the organs of perennation;as they can help the plant to tide over the unfavorable environmental conditions and then germinate/sprout to produce new plant/plants. This is equivalent to dormancy or hibernation in some animals as frogs and bears.

Perennating Bud (Perennating Organ)

The vegetative means hereby biennial and perennial plants survive periods of unfavourable conditions. The aerial parts die back to a minimum at the onset of unfavourable conditions, and food for the new shoots of the next growing season is stored in underground organs (e.g.tubers, bulbs, rhizomes), or in buds on the stems of woody plants. Seeds may also be considered perennating organs.

Mechanism of Persistance of Weeds

Weeds are highly persistent category of plants. They existed even before the first seed of any crop was planted on earth and they are with us even today, probably in larger number and with greater vigour. This speaks ample of their highly persistent nature. If a mixed growth of a crop and weeds is left to nature, soon there will be weeds all over, with no trace of the crop plants. High persistence of weeds results from their multifaceted mechanisms. Important among these are :

Prolific /Higher Seed Production Rate

Most weeds are prolific seed producers. For instance, the per plant seed production capacity of *Cuscuta* spp. was found approximately 16,000; *Conyza* 33,992; *Chenopodium album*, 72,000; and *Amaranthus*, 196,000. Only in few instances of some perennial weeds, the seed production was weak. Further, the immediate viability of weed seeds has been found to vary from 6 to 78%. Thus, one can anticipate that in majority of cases even if a few weedy plants escaped control measures in the field, they were sufficient to produce enough seeds to continue their progeny for years to come. It is further interesting to note that weeds can set viable seeds even when they are harvested before full maturity.

Table 1.2.1 Seed production capacity of some common weeds

Weed species	Average number of seeds /plant	Immediate germination percentage
Amaranthus spp.	1,96,000	Not recorded
Chenapodium album	72,000	Not recorded
Brassica nigra	58,000	Not recorded
Commelina benghalensis	2,450	27
Cynodon dactylon	170	6
Cyperus esculantus	820	38
Cyperus rotundus	40	0
Datura stramonium	13,900	5
Eluesine indica	41,200	61
Ipomea hispida	338	Not recorded
Physalis angulata	22,000	6
Portulaca spp.	193,000	6

Dormancy of Weed Seeds and other Propagules in Soil

As stated earlier, the majority of weed seeds lying below about 5 cm soil depth remain dormant, and act as source for future flushes of weeds. Permanent experiments laid out for up to 10 decades have shown that depending upon the weed species, the weed seeds retained viability for 2 to 100 years, to different levels. Perhaps, the oldest seed burial experiments were laid out in 1879 by late Dr. W. J. Beal in USA. The latest viability tests from these experiments were reported by Kiviban and Bavdurski (1973).

Vegetative Propagation

Many weeds are extremely persistent because of their ability to propagate by vegetative means. When the above ground parts of such weeds are destroyed, their deeply placed vegetative propagules put forth new shoots as soon as the external stresses are removed and favourable conditions revive.

Inherent Hardiness

To be widely successful a weed species must adapt itself to diverse environmental conditions. Weeds seem to possess some kind of built-in mechanism to survive against the vagaries of nature, like extreme cold, heat, drought, biotic stress, and soil abnormalities. This is inherent hardiness of weeds. It is hard to find weeds damaged under a night's severe frost, or destroyed because of attack by some insect pest, disease organism, or nematode. Weeds are rarely seen to either develop chlorosis or any other mineral deficiency or toxicity symptoms, even though the soil may be actually abnormal in its mineral composition. Also, poor quality water declared unfit for irrigating crops, fails to injure weeds. In fact, in crop fields all abnormal features of soil and water help weeds to grow vigorously in spaces vacated by the stunted growth (or death) of the crop plants. The mechanisms involved in this inherent hardiness of weeds are not much known. Further, several weeds of tropical origin, e.g. *Cyperus* and *Amaranthus* spp., have the advantage of adopting C_4 pathway of CO_2 fixation, which is devoid of photorespiration.

Evasiveness

Many a weed is capable of evading destruction by animals and man because of their bitter taste, disagreeable odour, spiny nature, and mimicry.

Self-Regeneration

Weeds are self-sown plants. They don't require any artificial, friable seedbeds for their germination. Detached from the mother plants, weed seeds and other propagules germinate profusely on undisturbed soils whenever the environment is favourable for the purpose.

Weed Succession

In nature, the individuals of a weed species often have chance to cross breed to variable levels. This leads to the development of a few plants of different genetic make-up, forming new races within a species. Such races of weeds are called **agricultural ecotypes.** When herbicides (or bio agents) are used continuously to destroy the "normal" races of a weed species, their newly developed ecotypes may sometimes prove tolerant to the herbicides used, when they get chance to gradually expand in numbers, each season. Such agricultural ecotypes are then called chemo types. The chemo types greatly aid a weed species to persist. Weed succession can also occur amongst different weed species themselves in response to long term adoption of an agricultural practice, including the use of herbicides. This leads to the destruction of the susceptible group of weed species, leaving behind few plants of the resistant species to gradually build up their population and finally emerge as the dominant weed flora of the area. In India, several examples of this kind have been recorded in recent years. Singh *et al. (1993)* reported that in the Punjab, in wheat fields, population of *Avena fatua* 'wild oat' was on increase because of several years' herbicidal control of *Phalaris minor*. Likewise, in paddy fields, the continuous use of butachlor has led to increase in the tolerant weed, *Ischaemum regosum,* wrinkle grass'. In Tamil Nadu, the repeated use of butachlor in paddy fields has led to increase in *Cyperus rotundus*, 'nut sedge' and some other perennial weeds.

Dormancy of Weed Seeds

Seed Dormancy

Dormancy is a state of seeds and buds in which they are alive but not germinated. Weed seeds possess a variety of special germination mechanisms adapted to changes in temperature, moisture, aeration, exposure to light, depth of burial of seeds etc., when conditions are unfavourable for germination, they can remain dormant or delay germination. If all weed seeds were to germinate at one time, their seedlings could be destroyed. Dormancy allows storage of millions of weed seeds in soil and enables them to grow in flushes over years. In this context, the old gardeners saying "*One year Seeding seven years Weeding*" is very appropriate. In fact, weed seeds have been found viable even after 20-80 years of burial in soil. Seeds of weed species of Boraginaceae, Convolvulaceae, Cucurbitaceae, Leguminoseae and Gramineae families have along dormancy period often running into several years.

Three types of dormancy are recognised: innate, induced and enforced.

1. Innate Dormancy

Innate dormancy is a genetically controlled character and it is a feature of specific weed seeds. Innate dormant seeds will fail to germinate even if they were present in the top 3 to 5 cm soil and adequate soil moisture and temperature conditions were provided to them. Innate dormancy usually results for reasons of either hard seed coats, like in *Setaria, Ipomoea,* and *Xanthium* spp.,or immature embryos, as observed in *Polygonum, Juncus,* and *Eleocharis* spp. In certain weed seeds, particularly those of xerophytic origin, the presence of specific germination inhibitors is responsible for their innate dormancy. In nature the innate dormancy of weed seeds is overcome with either passage of time, or under the influence of some climatic pressure.

The type of dormancy is characterised by (i) rudimentory embryos (ii) physiologically immature embryos resulting from inactive enzyme systems, (m) mechanically resistant seed coats or scales which prevent embryo expansion

(iv) impermeable seed coats preventing entry of water and/or gases, and (v) excessive presence of inhibitors.

i) Rudimentory embryos: The outside of the seed may appear fully developed, but it may have an immature embryo that needs more growth before the seed can germinate. Therefore, the seed appears dormant, although the embryo is slowly growing and developing. Seeds of *Typha latifolia* and *Polygonum* sp. show this characteristic.

ii) Physiologically immature embryos: In some species, the embryo appears completely developed, but the seed will not germinate although the seed coat has been carefully removed to permit easy absorption of water and oxygen. Light and darkness have no effect. In this case germination occurs normally after a period of after ripening. Occasionally cool temperatures for several months will end this type of dormancy. After ripening it is especially common in the grass, mustard, *Polygonum* sp. families.

After-ripening is a physiological change of a complex nature. Although the exact processes are not completely understood, they may be associated with changes in the storage materials present, substances promoting germination may appear, or substances inhibiting germination may disappear (Ashton & Monaco, 1991).

iii) Mechanically resistant seed coat: Seeds of few weed species may fail to germinate even though water and air can get inbecause of the mechanical resistance of the seed coat which may withstand pressure of 100psi (pounds per square inch). The seed will quickly germinate if the seed coat is removed. *Amaranthus retroflexus* and *Brassica nigra* have this type of dormancy. As long as the seed coat remains moist, it remains tough and leathery for as long as 50 years. Any factor that weakens the seed coat will help break dormancy. Drying at temperature of 110°F or mechanical or chemical injury to the seed coat may break this type of dormancy.

iv) Seed coat impermeable to water/air: The prevention of water uptake into the seed that has an impermeable seed coat is a very effective dormancy mechanism occurring in weeds which are members of several families, particularly those of Leguminoseae, Malvaceae, Chenopodiaceae and Solanaceae (Werker, 1980). Many of the most troublesome and long-lived weed seeds have coats that are impermeable to water. These seeds, also termed as hard seeds, respond to scarification of the seed coat for imbibing water and germinating. Alternative cycles of temperature or hydration have been suggested as natural means of increasing permeability of seed coats to water.

v) Inhibitors: Inhibiting chemicals in the fruit or seed coats, or within the embryo or embryo, may delay germination. For example, the fruits of tomato and cucumber and the coats of lettuce and beet contain substances which prevent germination. This is also true for many fruit trees. In the soil, these inhibiting materials diffuse away from the seed or are absorbed by the soil particles, until they reach a level at which they are ineffective. Their period of effectiveness depends on their stability or on the prevention of diffusion out of the seeds by semi permeable membranes in either the living protoplasm or nonliving seed coat. Some chemicals like cumarin, ferulic acid and abscisic acid (ABA) etc. are present in the seeds of weeds which do not allow these to germinate. The seeds will germinate only when the concentration of these inhibitors come down.

Dormancy can be broken by the following processes

- Treating the seed with anti dormants like Gibberellic acid (GA) and it is only possible under lab conditions.
- The seed coat decays with passage of time and it becomes permeable to water and oxygen.
- Inhibitors are also washed away with rain or irrigation water and their effects are diluted.
- With time slow chemical changes in inhibitors also takes place and seed show germination under favourable conditions.

2. Induced Dormancy

Seeds which ordinarily would germinate immediately if planted under favourable conditions may be thrown into dormancy by an unfavourable environment so that they will not germinate even when conditions become favourable. This is termed secondary or induced dormancy.

Induced dormancy results from some physiological change in otherwise non-dormant weed seeds under the impact of factors like a marked rise in soil temperature, increased CO_2 content of the soil, low O_2 pressures, water logging etc. certain weed seeds, like those of wild oat, exhibit all the three kind of dormancies.

3. Enforced dormancy

Enforced dormancy in weed seeds is due to their placement deeper than 5 cm, resulting usually from tillage of the field. Weed seeds under this kind of dormancy germinate readily whenever these are restored to the top 3 to 5 cm layer of soil by tillage, provided adequate soil moisture and congenial temperatures were available in this zone of the soil. Enforced dormancy is a non-specific character of the seeds and it is caused by the absence of red light (r) under the ground, which otherwise induces germination in seeds by activating their phytochrome system (P), comprising a responsive chromophore blue pigment attached to the protein molecule in seeds. Far-red light (fr) deactivates the system and thus induces dormancy in seeds.

$$Pr < \frac{\text{far red}}{\text{red}} Pfr$$

Cultivation counters enforced dormancy by bringing the weeds to surface where they are exposed to full sun-light spectrum, besides better aeration. Higher soil temperature and NO_3 content of surface soil may further help in breaking the enforced seed dormancy.

Allelopathy

Many plants release some chemicals called allele chemicals during different growth stages which may affect the growth and development of adjoining plants. The phenomenon of one plant having detrimental or promotory effect on another plant through the production or release of toxic chemicals has been termed allelopathy. These allele chemicals may have synergestic or antagonistic or neutral effects. For instance mustard plants release allele chemicals which stimulate the germination of parasitic weed like *Orobanche aegyptiaca*.

Importance of Weed Biology in Weed Control

The effective use of control measures, whether cultural or chemical, requires an appreciation of the biological characteristics of weeds concerned. Correct identification of the species present is the first step, before appropriate control methods could be selected.

The time during the year at which germination occurs is important for deciding weed management strategies which may be chemical, mechanical or cultural ones. Where thegermination period is restricted, it may be possible to take advantage of this in planning the crop sequence so that competition during the early stages of crop growth is avoided. For example, in northern India, *Phalaris minor* starts germinating in beginning of November with falling down of temperature. If wheat is sown earlier i.e. in last week of October, crop can develop earlier ground cover and suppress the growth of *Phalaris minor*. Alternatively early flushes of *Phalaris minor* could be killed by using stale-seed-bed technique and then sowing the crop afterwards. In stale bed, sowing of wheat after pre-sowing irrigation is delayed so that maximum weed seeds can germinate and these seedlings are killed with contact herbicide or by ploughing the field. Later flushes do not compete that much with the crop and also could be killed by using herbicides. Similarly in rice nursery time of germination of *Echinochloa crusgalli* is the same. If nursery could be kept free of weeds either by applying herbicides or by physical methods, weed competition in transplanted crop could be avoided.

Periodicity of germination of weeds must be taken into consideration in deciding when to apply herbicides. It is useless to apply a relatively non-persistent pre-emergence treatment or a contact herbicide before the species to be controlled has begun to germinate, and too early application during the germination period may be followed by the emergence of sufficient undamaged seedlings to constitute a weed problem. Depth of germination is an important consideration while planning cultivations which aims at depleting the weed seed content of the soil by stimulating germination and destroying the resulting seedlings. The first cultivation of a series should always be the deepest.

Images of some commonly grown weeds

Amaranthus spinosus

Amaranthus viridis

Ammania bacifera

Anagallis arvensis

Aristalochia bracteata

Asphodelus tenuifolius

Jetropha gossypifolia

Lathyrus aphaca

Lippia nodiflora

Commelina subulata

Chorchorus olitorius

Chorchorus capsularis

Chenapodium album

Coccinia indica

Commelina benghalensis

Eclipta alba

Eleusine indica

Acanthospermum hispidum

Achyranthus aspera

Aerva tomentosa

Aerva lanata

Tribulus terrestris

Trichoderma indicum

Dtura factuosa

Digera arvensis

Cyperus difformis

Dactyloctenium aegyptium

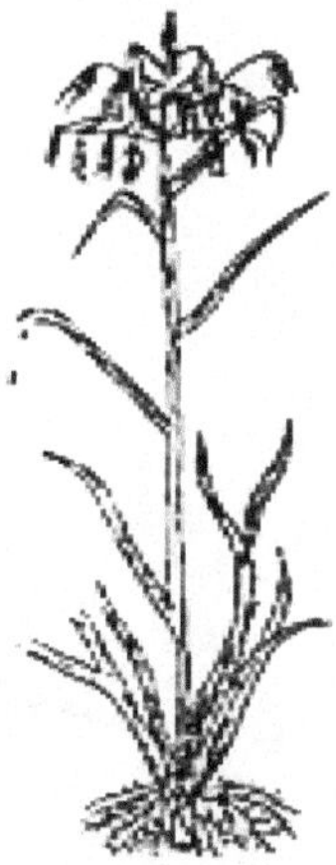

Avena fatua

Bidens pilosa

Boerhaavia diffusa

Ludwigia parviflora

Panicum repens

Merremia emerginata

Fimbristylis miliacea

Flavaria australasica

Phalaris minor

Phaseolus trilobus

Physalis minima

Paspalidium germinatum

Pavonia zeylanica

Perotis indica

Tridax procumbens

Typha angustata

Vallisnaria spiralis

Heteropogon contortus

Iseilema iaxum

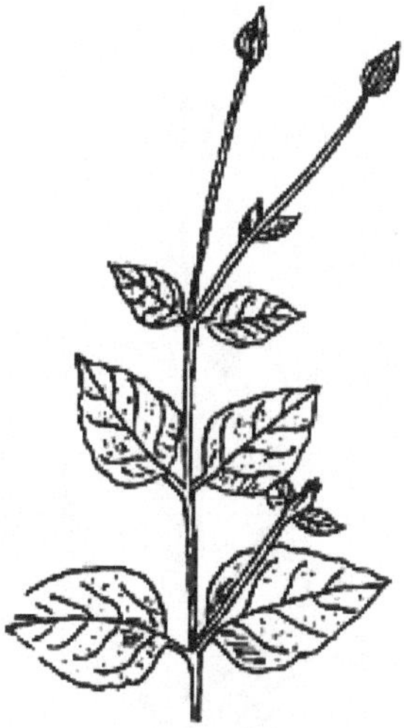

Spilanthes calva

Tephrosia purpurea

Pistia stratioites

Rungia repens

Solanum xanthocarpum

Calotropis gigantea

Cenchrus setigerus

Chavorophora rotteleri

1.3

Weed Ecology

Weed Ecology

Ecology is the study of reciprocal relationship between organisms and their environments. The ecology of weed is thus concerned with growth characteristics and adaptations that enable weeds to exploit the ecological niches left open in those environments altered by man for his use and the survival mechanisms where by weeds persist in such environments. In simple words weed ecology is reciprocal relationship of weeds with their environments. Weeds grow voluntarily in all crops raised by human beings since ancient ecology of weeds is divided into two categories (types).

1. Weed Auto Ecology

It is study of only one population i.e. how an individual species in a population interacts with its environment at all stages of its life cycle. So, in case of auto ecology competition within same species (intraspecific) or with crop plants (inter specific competition) is studied.

2. Weed Synecology

The study of relationships between communities of different weed species with their environment is called weed synecology. Different weed species grow in certain combinations and in specific patterns in the ûeld. The population of different weed species interact with the external environmental conditions and among themselves. Weed communities differ from field to field and crop to crop in the same locality. The community character is dynamic and changes in form and composition of weed flora takes place in course of time due to some external and internal factors prevailing in the community. This process is called ecological succession. With the passage of time there will be series of changes in the composition of weed community which ultimately results in stable community. This is termed as climax community.

Ecological Succession of Weeds

The green vegetation that initially invaded the bare rock is slowly replaced by second type, which itself is replaced by third type in course of time. Weed communities that replace one another in a orderly sequence over the years is called as ecological succession of weeds.

Crop Weed Association

For successful cultivation of various crops, optimum climatic and edaphic conditions are required. A particular community of weeds whose ecological requirements tally with the crop are generally found to be associated. *Asphodelu stenuifolius* (Piazi/wild onion) and *Carthamus oxyacantha* (Pohli) have low water as well as fertility requirements and these weeds were very troublesome weeds of rainfed crops like wheat, barley and gram at sometimes (during early decades). But with the introduction of dwarf wheat varieties having high requirements for irrigation water and nutrients, these weeds (*Asphodelu stenuifolius* and *Carthamus oxycantha*) do not appear under such conditions as this dwelling is not suitable for their growth. The rice-wheat rotation gained more popularity during the past many years. So instead of low water/fertilizer requiring weeds, the infestation of *Phalaris minor* (Gullidanda/Sitti) became more prominent as the prevailing growing conditions were favourable for this weed to grow. The puddling as well as water stagnation conditions which are pre-requisite for rice are ecologically favourable for survival, growth and development of Phalaris *minor*. The seeds of *P. minor* are capable of tolerating anaerobic conditions by entering into secondary dormancy, which perhaps is also one of the reasons of its better adoption in rice-wheat cropping system (Malik *et al.*, 1998).

Morphological Similarities

Morphological similar weeds are difficult to control both by mechanical or chemical methods. For example plants of *Echinochloa crusgalli* (Swank) which resemble with rice are often uprooted with rice nursery and well grown plants of this weed when transplanted with rice seedlings are not controlled with any of the herbicide recommended for this crop. These weed plants thus produce lot of seeds and increase soil seed bank for the following years. Similarly plants of *Phalaris minor* and *Avena ludoviciana* are not controlled (particularly intra row plants) with mechanical methods. Seeds of *Cichorium intybus* (Kashni) have resemblance with seeds of berseem, so, this weed makes its association with only berseem crop, due to similarities of seed of both plants.

Seed Shedding Behaviour

Majority of weed plants complete their life cycle before the harvest/maturity of the crop(s) in which they are associated. The seeds which are shed ill the field during one season, become the source of infestation during the coming years.

About 90 to 95 per cent seeds of problematic weeds like *Phalaris minor, Avena ludoviciana, Echinochloa crusgalli, Ischaemum rugosum* and *Trianthema portulacastrum* are shed in the field. This seed shedding behaviour is one of the main reason of association of a particular weed with a specific crop when sown on the same field regularly. On the other hand weeds with delayed maturity (like *Lolium temulentum*) than the crop, loose their association, however, use of infested wheat seeds is the only reason of association of *Lolium temulentum* with wheat crop.

Escape Removal Through Sieving

The seeds of few weed seeds is of the same size as of the crop grains/seeds and thesecannot be separated out even by thorough sieving. These seeds are sown in the ûeld along with crop seeds. The size of few weed seeds like *Convolvulus arvensis* (Lehli), *Rumex spinosus* (Kandiali Palak), *Avena ludoviciana* (Wild oats) etc. are of same size as of wheat grains and these cannot be removed even with thorough sieving. The crop infested with these weeds should not be kept for seed production purposes.

Congenial Environment

The association of weeds not only depends upon crop but it is also influenced by cropping sequence. For example field having infestation of wild oats in wheat crop can be eliminated if brought under rice-wheat rotation for 2-3 years. The wild oat seeds being porous absorb water and loose viability with continuous standing of water. The continuous wet conditions eliminate wild oats but may give favour to *Phalaris minor* establishment due to congenial environmental conditions created with the cultivation of paddy.

Germination in Flushes

The weed seeds which have the tendency to germinate in many flushes, are very difficult to control both with mechanical or chemical methods. The flush of weeds which appear after the adoption of weed control measures, produce seed and such type of weeds are rather permanently associated with the crop in spite of adoption of best weed management practices. During *kharif* season, *Trianthema portulacastrum* (Itsit/Chupati) has the tendency to germinate in many flushes from April to October in all crops excepting rice. The plants of this weed have a tendency to set seeds even if they appear during latter stages of crop growth or late in the season. By taking the advantage of this character, *Trianthemaportulacastrum* weed is associated with all *kharif* crops.

Continuous use of One Group of Herbicide

The continuous use of single group of herbicides over years, leads to resistance phenomenon. For example, with continuous use of isoproturon in wheat, *Phalaris minor* had developed resistance to isoproturon in Punjab (Walia *et al.*, 1997). In spite of use of isoproturon even at higher doses i.e. 3 to 4 times than the recommended with proper spraying technology, *Phalaris minor* was not killed. So, this weed continued to associate wheat through physiological and genetic modifications. Crop rotation as well rotational use of herbicides, helps a lot to avoid association of a particular weed with a particular crop.

Allelopathic Effects

There are few weeds which are only associated with a particular host. In the absence of host plant, these weed seeds show no germination even when the seeds are in viable conditions. For example *Striga densiflora* is a parasitic weed of sugarcane and *Orobanche aegyptiaca* is of mustard crops. These host plants (crops) excrete some chemicals in soil which are called allele chemicals and these chemicals acts as stimulants for the germination of seeds of these parasitic weeds which are associated with their speciûc host crops while seeds of parasitic weeds do not germinate when other crops are sown in their habitat. Bhowmik and Doll (1992) reported soybean yield reduction ranging from l4 to 19 per cent by the extracts of dried residues of several annual species: *Chenopodium album, Amaranthus retroflexus, Abutilon theophrasti and Helianthus annuus* (Sunflower).

The occurrence of a weed in a particular locality depends upon many factors and the important are :

1. Physiographic
2. Edaphic
3. Climatic

Physiographic Factors

These factors relates to the form and behaviour of the earth's surface and the distribution of weed plants is highly dependent on physiographic factors which includes geology of soil, topography, altitude, exposure to sunlight as well as direction of mountains etc.

On slopy soil, water flows down before it has time to be soaked into the soil. The amount of rainfall lost by runoff increases with the steepness of the soil and the water absorption decreases. Therefore, in spite of heavy rainfall, a sloping soil is usually dry. Hence, xerophytic plants like *Euphorbia agave* and

E. arboreous are met on slopes of western Himalayas. However, due to more water holding capacity of plain soils (valleys) its weed flora is different than of mountains. Further. there is significant change in vegetation with the drastic change in the environmental conditions due to increase in altitude.

A slope exposed to sun and wind has vegetation different from its counterpart. The southern slopes of Himalayas has mesophilous and the northern slopes have xerophilous vegetation, because southern slopes come across monsoon winds and are humid but the northern slopes are dry due to dry winds.

Edaphic and Climatic Factors for Weed Occurrence

Edaphic Factors

The role of soil fertility in manipulating weed-crop competition has been investigated extensively in the hope of reducing it to the advantage of the crop. But, unfortunately, most experimental results have not favoured this. The elevated soil fertility usually stimulated weeds more than the crop, reducing thus the crop yields. It then derives that improving soil fertility, without controlling the weeds, strengthened the competing ability of weeds for other plant growth elements like soil moisture and light although the competition for the applied fertilised, simultaneously. Elevation of soil fertility of a weedy field resulted in an uneconomic and very low fertilizer use efficiency.

But variable results when fertilization improved the growth of a weedy crop and suppressed the weeds, are also available. Such results are obtained when the soil was acutely deficient in a particular nutrient, and the nutrient was applied selectively, both in time and space, so that it was largely available to the crop plants.

The moisture variations in soil may also influence the course of weed-crop competition. Weeds are adapted to grow well and compete with the crops, both in moisture stress and ample moisture conditions. This is so because there are different species of weeds available suited to highly variable moisture conditions. Removal of an intense moisture stress by irrigation may thus benefit crops more than the weeds, leading to increased crop yields. However, it all depends upon the weed-crop situation at the time of irrigation. If the weeds were already present at the time of irrigation, they would often grow so luxuriantly as to completely over power the crop. But if the crop is irrigated after it has grown 15 cm or more tall in a weed free environment, irrigation could hasten closing-in of crop rows, thus suppressing the weeds. In any case, amelioration of weed-crop competition by irrigation is no substitute to weeding.

In submergence conditions of paddy cultivation, weeds are put to a disadvantage to start with. But if there is a break in submergence, the weeds may germinate and grow more vigorously than the crop, even if fields were submerged later on. Our knowledge of weed-crop competition for moisture is very much limited.

Abnormal soil reactions (very high or very low pH) often aggravate weed competition. It is because while specific weed species suited to different soil reactions exist with us, our crops grow best only in a specified range of soil pH. As such, weeds offer more intense competition to crops on abnormal pH soils than on the normal pH soils. No critical studies on this aspect of weed competition are, however, available on hand.

Soil is the top cover of the earth in which plants can grow. It is made up of a complex composite mass of weathered rock and minerals, decomposed organic matter of wild plants. Soil provides support, mineral nutrition, water supply etc. Different types of weed species grow in different soils. Soil fertility, structure and its other physical and chemical conditions have a great impact on the distribution and establishment of weed flora. Depending upon soil type/conditions, weeds are divided into following groups.

Weeds of High Fertility Soils

Certain types of weeds require a dwelling of good fertile lands with high organic matter for their proper growth and development. For example- *Phalaris minor, Eleusine aegyptiacum, Cirsium arvense, Trianthema portulacastrum, Amaranthus viridus, Convolvulus arvensis, Echinoch loacrusgalli*, etc.

Weeds of Low Fertility Soils

For growth and development of some weeds, low fertility lands with low organic matter content are required. Examples are *Carthamus oxyacantha, Asphodelus tenuifolius, Celosia argentia, Mullugoverticellata, Sileneconodia, Saccharum spontaneum, Ipomoea pestigridis*. These weeds prefers to grow on soils with more sand content as these soils retain less water which is prime requirement for their growth.

Weeds of Alkaline Soils

Few weeds can grow well on high pH soils in which all crops cannot grow well. For example *Anagallis arvensis, Stellaria media, Polypogon* spp., *Lophochloa* spps.

On Acidic Soils

Euphorbia spp., *Spergula arvensis* are examples of weeds which can grow well on low pH soils.

On Swampy Soils

Some weeds can grow on soils with very high water holding capacity. *Poaannua* and *Cirsium arvense* are the examples of *Rabi* season weeds whereas *Echinochloa crusgalli* and *Cyperusiria* are weeds of *kharif* season.

Climatic Factors

Climate influences the species composition and density of weeds. Therefore, it cannot be without its influence on weed-crop competition. Adverse weather conditions e.g. drought, excessive rains, and extremes of temperature, will favour weeds since most of our crop plants are susceptible to climatic stresses. The weed competition is further intensified when crop cultivation is stretched to marginal lands and too far from their optimum climatic requirements. All such stresses weaken crop's inherent capacity to fight weeds.

Further, the time of germination of weeds in relation to crop germination is a primary factor in weed-crop competition, which is governed by climate. And our knowledge of the exact germination requirements of weeds is so meagre that we are unable to anticipate their germination behaviour.

Scanty literature available on factors affecting the germination of weeds in field conditions shows that weed emergence in fields was more rapid on soils kept continuously wet by rainfall than those that dried three days after planting crops. The differences in germination of weeds narrowed as the temperature increased, because then the weed seeds absorbed enough moisture in the first three days wet period itself before the soil dried. For germination of seeds of *Amaranthus retrofiexus* 'redroot pigweed' and *Echinochloa colonum* 'barnyard grass' best temperature was 17°C for eight hours at night and 25°C for 16 hours in the day.

Temperature, moisture and light are very important climatic factors on which distribution and adaptation of weed plants depends.

Temperature

Both soil and above ground temperature are important which govern reproduction and establishment of weeds. Weeds can be grouped into two categories depending upon their requirements for temperature.

a) ***Rabi* season weeds:**The seeds of *Rabi* season weeds germinates when temperature falls in between 10 to 20°C. For example *Phalaris minor, Avena ludoviciana, Chenopodium album* etc.

b) ***Kharif*season weeds:** The seeds of these weeds show germination only under high temperature condition i.e. when temperature exceeds 20°C,

Trianthema portulacastrum, Echinochloa crusgalli, Eragrostisspps. etc., are *kharif* season weeds.

The seeds of some weeds like *Avenaludoviciana*, do not loose their viability even if they remain under snow for a period of 3-4 months whereas *Cynodon dactylon, Cyperusrotundus* and *Sorghum halepense* cannot thrive when the soils are exposed to frost.

Moisture

Soil moisture received through rainfall or irrigations plays a key role on the distribution of weeds. Depending upon moisture requirements weeds can be classified into three groups.

a) **Hydrophytes** : These are aquatic plants which normally grow in water and these are characterised by sponginous of tissues e.g. *Lemna minor, Eichhornia crassipes, Potamegeton* sp. etc. If water is not available these weeds will die.

b) **Mesophytes** : These are the weeds of arable crops and their moisture requirements are same as of crops. These weeds will wilt permanently after loosing 25 to 50 percent of their total water content, e.g. *Phalaris minor, Avena ludoviciania, Trianthema portulacastrum, Eleusine aegyptiacum, Portulaca oleracea* etc.

c) **Xerophytes**: These weeds have low water requirements and can tolerate prolonged drought without any adverse effect on growth. These weeds show wilting only after loosing 50 to 75 per cent water content, e.g. *Zizyphus rotundifolia, Calotropis procera, Aervapersica, Tribulusterrestris* etc.

Light

The intensity and duration of light plays very important role in plant distribution. Duration of light (day length) is very important for flowering and seed setting. Depending upon light requirements, weeds can be grouped into two categories.

a) **Sciophytes**: Some weeds can grow/tolerate reduced light conditions and these are called as sciophytic weeds e.g. *Solanum nigrum, Agropyron repens, Ageratum conozoides*.

b) **Heliophytes**: Weeds requiring bright sunlight conditions for their growth and development e.g. *Phalaris minor, Celosia argentea, Cyperus rotundus* etc.

Crop-weed Competition

Competition is the action of endeavouring to gain what another endeavours to gain at the same time. In other words, competition is mutual adverse effect of organisms utilizing common resources which are in short supply. Among plant communities each plant is in a state of continuous competition with its neighbouring plants for various growth elements, both above and under the ground. Competition between crop and weeds is mainly for nutrients, soil moisture, light and space etc. The basic requirements of crop plants and weed plants are the same and competition will start only when the supply of any one or more factor(s) is in short supply. If crop plants occupy the soil and are vigorous, weeds are excluded or retarded in their growth. On the other hand if the crop stand is thin or lacks vigour, weeds will flourish. Our concern here is with such a competition between the weeds and the crops growing on agricultural lands which tends to limit, or even extinct, the weaker competitor; invariably the crops.

There are two types of competition which exist in field crops/weeds.

1. Intra Specific Competition

It is the competition between the plants of same species which may be of crop or weed. As the plants (of crop and weed) with time, due to mutual competition, their number/unit area decreases which is due to self induced competition. For instance, competition within plants of *P. minor* is called intra specific competition. Also, the competition between the plants of same crop is also termed as intra specific competition.

2. Interspecific Competition

It is competition between two or more plants belonging to different species. In other words, inter speciûc competition is competition of crop with associated weeds and it has more bearing on agricultural crops. Competition within weeds belonging to different species (*P. minor* x Wild oats) is also included in this category. Aggressive species may dominate in competition due to early germinating habits, quick growing rates during initial growth stages.

Tall growing habits, more leaf area, extensive root system, more spreading habits and high seed production rates.

Weeds compete with crops for water, nutrients and light. Being hardy and vigorous in growth habit, they grow faster than crops and consume large amount of water and nutrients, thus causing heavy losses in yields. An estimate shows that weeds can deprive the crops 47% N, 42% P, 50% K, 39% Ca and 24% Mg of their nutrient uptake.

Crop	Reduction in yields due to weeds (%)	Crop	Reduction in yields due to weeds (%)
Rice	41.6	Groundnut	33.8
Wheat	16.0	Sugarcane	34.2
Maize	39.8	Sweet potato	70.3
Soybean	30.5	Carrot	47.5
Gram	11.6	Cotton	72.5
Pea	32.9	Onion	68.0
Millets	29.5	Potato	20.1

Competition for Nutrients

Competition for nutrients constitutes an important aspect of weed-crop competition. While N has been investigated extensively and it may be the most important element inplant competition, but the competition may occur also for other nutrients required for plant growth. Weeds usually absorb mineral nutrients faster than many of our crop plants and accumulate them in their tissues in relatively larger amounts. Competition will depend upon type and density of weed flora associated with the crop. Generally, morphologically similar weeds are more competitive than morphologically dissimilar weeds and depending upon their intensity and type of crop, these weeds can trap 30% to 60% of the applied nutrients due to their quick growing habits (Walia and Gill, 1985).

Similarly, Gupta (1998) found that from a weedy sugarcane field, weeds removed about 162 kg N, 24 kg P_2O_5 and 203 kg K_2O/ha against only 28 kg N, 17 kg P2O5 and 62 kg K_2O/ha by the crop. Weeding the field elevated the nutrient removal by sugarcane to 111 kg N, 81 kg P_2O_5 and 253 kg K_2O per ha. The potato field lost on an average 63 kg N, 11 kg P2O5 and 88 kg K_2O/ha through weeds (Mani, 1975). Singh and Kolar (1994) reported that a weedy crop of soybean could remove only 87.7 kg N/ha against a weed free crop which removed 132.7 kg/ha. Shahi (1978) estimated a loss of 32 kg N/ha each year through *Chenopodium album* alone in dry land conditions.

Competition for Moisture

Moisture is the most precious as well as the basic input for germination and growth of plans. Research has established that weed-crop competition for moisture usually occurs alongwith other elements of competition. It becomes increasingly critical with increasing soil moisture stress, as occurring in the dry farming areas of the semi-arid and arid regions of the world. In general, for producing equal dry matter, weeds transpire more water than do most of our crop plants (Table 1.3.3). Therefore, in dryland agriculture the actual evapotranspiration from the weedy crop fields is much more than the evapotranspiration from a weed free crop field. In such an agrisystem, during a dry spell the weedy crops exhibit wilting or other moisture stress symptoms much earlier than a weed free crop. So much so, in a weedy field the soil moisture may be exhausted much before the cropreaches its fruiting stage, which is often the peak consumptive use period of the crop, causing significant cuts in crop yields.

Table 1.3.2: Nutrient Removal by Weeds and Crop Under Weedy and Weed-Free Situations*

Treatment	Nutrient removal (kg ha^{-1})					
	By Weeds			By Crop		
	N	P_2O_5	K_2O	N	P_2O_5	K_2O
Wheat						
Unweeded	21.7	2.9	28.2	59.3	12.8	46.8
Weeded	1.2	0.2	1.6	75.9	15.6	69.2
Paddy						
(Transplanted)	74.0					
Unweeded	0	6.0	108.0	60	5	9
Weeded		0	0	174	14	238
Maize	46.7					
Unweeded	6.1	7.1	61.6	46.4	6.3	82.5
Weeded		0.9	8.0	143.7	21.7	190.9
Groundnut	163.0					
Unweeded	10.1	21.9	142.0	55.2	21.2	27.2
Weeded	162.2	2.7	9.2	150.6	60.6	74.4
Sugarcane	49.7					
Unweeded		23.8	202.9	28.3	17.3	61.8
Weeded		7.7	110.6	110.6	80.7	253.1
Mentha (1st	59.6					
harvest)	0					
Unweeded		13.0	77	24.2	5.2	34.6
Weeded		0	0	100.0	13.0	104.0

*Compiled from different sources cited in the bibliography

Similar observations were reported by Wiese and Vandiver (1970). In irrigated fields too, weeds grab lion's share of the applied water. Saini *et al.* (1993) found that in a weedysituation irrigation increased the sugarcane yields by less than 1t/ha in 1984, and by lessthan 3t/ha in 1985 against about 10 and 28t/ha increases in the weedfree plots in the two years, respectively (Table 1.3.5).

Table 1.3.3: Transpiration Coefficients of Certain Weeds and Crops (After Kanitkar*et al.*, (1960)

Plant species	**Transpiration coefficient 'Q'**	**Plant species**	**Transpiration coefficient 'Q'**
WEEDS		*Tephrosia purpurea*	1108
Amaranthus viridis	336	*Tribulus terrestris*	221
Cynodon dactylon	813	*Tridax procumbens*	1402
Digitaria sanguinalis	696	*Xanthium strumarium*	331
Echinochloa colonum	674	CROPS	
Ischaemum pilosum	556	*Zea mays*	352
Salsola kali vartenuifoila	542	*Sorghum vulgare*	394 (430)

Table 1.3.4: Rooting Depths of Some Common weeds (After Rao, I. M. 1961. paper presented at the 1[st]All India Weed Control Seminar held. Oct. 3-4, at Bombay, India)

Weeds species	**Maximum observed rooting depth (m)**	**Weeds species**	**Maximum observed rooting depth (m)**
Amaranthus paniculatus	0.5	*H. indica*	1.0
Cleome sp.	0.6	*Polygonum p/ebejum*	1.2
Convolvulus arvensis	2.4	*Portulaca oleracea*	0.8
Eclipta alba	0.9	*Trianthema monogyna*	1.2
Euphorbia prostrata	1.3	*Tribulus terrestris*	1.2
Heliotropium eichwaldii	1.0	*Voluterlladi varicata*	2.3

Table 1.3.5: Effect of Irrigation on Sugarcane Yield Under Variable Weedy Situations

Treatments (weed control)	**Sugarcane Yield (t/ha)**			
	1984-85		**1985-86**	
	Unirrigated	**Irrigated**	**Unirrigated**	**Irrigated**
Unweeded	23.85	24.70	24.93	27.59
Handweeded	37.45	47.16	38.42	66.89
Atrazine 2 kg/ha, PE	31.31	38.39	31.86	58.57
C. D. Interaction (P=0.05)	2.40	3.76		

Comments: In the unweeded situation irrigation could not increase the cane yield, while in theweedfree situation it increased it, markedly.

Competition for Light (Solar Energy)

Light is required by plant leaves for photosynthesis. The growth of weeds in a weedy crop will cut light penetration towards crop, hence reducing its photosynthetic efficiency. Few weeds like *Trianthema portulacastrum* are very quick growing and can smother the *kharif* season crops such as cotton, maize etc. while huge intensity of *Phalaris minor* can cover wheat crop. So, when weeds cover the crop, PAR interception by the crop is hindered and availability of light becomes limiting factor resulting into poor crop yields due to less photosynthetic rates.

Plant height and vertical leaf area distribution define effective components of the competitive struggle for light (Graf & Hill, 1992). It becomes most important element of weed-crop competition when moisture and nutrients in soil are plentiful, and weeds havean edge over the crop plants in respect of their height. Light competition may commence very early in the crop season if a dense weed growth smothers the crop seedlings. Crops like cotton, potato, several vegetables, sugarcane, etc. are prone to heavy weed growths at their seedling stages and, consequently, suffer badly at the hands of shading effects of weeds. An important feature of competition for light in plant communities is that unlike competition for N and moisture, once the crop seedlings are shaded by weeds, later on providing additional solar energy cannot make up for the crop stunting caused earlier.

Competition for Space

Depending upon the growing habits, row to row and plant to plant spacings are different for different crops. In wider row crops, sufficient space is available for growth and development of weeds and also in crops which takes more time for emergence. Few cropslike sugarcane, turmeric, potato which takes long time for emergence, weeds grow without any competition and sometimes over power the crop resulting in crop mortality apart from poor growth and development of crop. In order to reduce competition for space, crop geometry should be modified in such a way that minimum space is left at the disposal of the weeds to grow.

Critical Period of Crop-weed Competition

The critical period of weed-crop competition is deûned as that shortest time span in the ontogeny of a crop when weeding will result in the highest economic returns. In other words, the critical period is the period in the life cycle of the crop during which there is maximum loss in grain yield due to competition by weeds. The crop yield levels obtained by managing the weeds during this period should provide crop yields sufficiently close to those obtained by the full season freedom from weeds. In general, initial growth stages of crops are very sensitive to competition by weeds and there is maximum loss of the crop if weeds are not removed during this period. With passage of time crop establishes and the severity of loss due to weeds gets decreased. So, it is very essential to control weeds during initial an period which is most critical for crop-weed competition. These uncontrolled weeds cover the crop and sometimes may result in crop failure.

So, keeping crop free from weeds by any method helps in improving yield drastically. A fundamental principle of plant ecology is that early occupants of the ground tend to discourage or exclude the later ones. The weeds that germinate either before or along the crop offer stronger competition. Mishra (1997) reported critical period of crop-weed competition in different crops (Table 2.3.6).

Table 1.3.6: Critical period of crop weed competition in important field crops. (days)

Crops	Critical period (days after sowing)	% Reductionin grain yield
Cereals		
Rice (direct seeded)	15-45	15-90
Rice (transplanted)	30-45	15-40
Wheat	30-45	20-40
Maize	15-45	40-60
Sorghum	15-45	15-40
Pearl millet	30-45	15-60
Pulses		
Pigeonpea	15-60	20-40
Green gram	15-30	25-50
Black gram	15-30	30-50
Cow pea	15-30	15-30
Chickpea	30-60	15-25
Peas	30-45	20-30
Lentil	30-60	20-30
Soybean	20-45	40-60
Oilseeds		
Groundnut	40-60	40-50
Sunflower	30-45	30-60
Castor	30-50	30-35
Safflower	15-45	15-40
Sesamum	15-45	15-40
Rapeseed mustard	15-40	15-30
Linseed	20-45	30-40
Commercial crops		
Sugarcane	30-120	20-30
Potato	20-40	30-60
Cotton	15-60	40-50
Jute	30-45	50-80
Vegetable crops		
Cauliflower	30-45	50-60
Cabbage	30-45	50-60
Okra	15-30	40-50
Tomato	30-45	40-70
Onion	30-75	60-70

In sugarcane Singh *et al.* (1980) found 4 to 16 weeks' period after planting critical for competition from weeds. But in potato only 2 to 4 weeks' period after planting was considered crucial for weeding. Delayed weeding caused considerable shrinkage in tuber yield, (Gupta *et al.,* 1979). A rapidly growing crop like mungbean responded most to weeding 35 days after sowing (Singh *et al.,* 1978). In blackgram the critical period of weed-crop competition was 30-34 d.a.s. In sunflower it was reported between 20 to 40d.a.s.

Table 1.3.7: Effect of Three Wheat Genotypes on the Dry Matter Production of Associated Wild oat (*Avena fatua*) and Grain Yield of Wheat. Mean data of three years (After Sandhu *et al.*, 1981, Ind. J. Weed Sci. 13(2) 146-49).

Wheat genotype	Dry biomass of wild oat (q ha^{-1})	Grain yield of wheat (q ha^{-1})	
		Unweeded	Weeded
Tall genotype (C-306)	28.3	27.6	36.8
Two gene dwarf (WI-711)	40.1	22.6	46.6
Three gene dwarf (WL-903)	54.7	19.7	43.6

Weeds in small grain crops usually first tend to reduce their number of ear bearing shoots per unit area (ebt). But as the weed competition becomes more intense, the ear length and thousand grain weight of such crops are also adversely affected by the weeds.

Cultural Practices for Reducing Crop-Weed Competition

With the adoption of cultural practices/agronomic manipulations, crop-weed competition can be minimized.

Cropping practices, such as time and method of planting crops, crop density andgeometr, and crop species and varieties have pronounced effects on weed-crop interference.

Adjusting Dates of Sowing

In general, when the time of germination of a crop coincides with the emergence of first flush of weeds, it leads to intense weed-crop interference. Usually, more the delay in the emergence of weeds, less is the weed-crop interference. By adjusting date of sowing of a crop before the optimum germinating time of its trouble-some weed(s), weed-crop competition can be reduced significantly by way of escape mechanism. Kolar and Mehra (1992) reported that October sown wheat attracted weed biomass of 22.0 q/ha,against November and December planted crop which permitted weed yield of only 10-12 q/ha. It is due to sowing the wheat early during the recommended schedule will help in avoiding first flush of *Phalaris minor* or sowing wheat in December (late) help in reducing the population of all type of weeds.

Depth of Planting / Sowing

The method of planting crops may also affect the course of weed-crop competition incrop fields. Weed seeds germinate most readily from the top 1.25 cm of soil, though it is considerable up to 2.5 cm depth (Yenish, *et. al.*, 1992). Only few weeds such as, *Avena, centaurea, Fumaria, Galium, Vicia*, and *Xanthium* spp., germinate from even 15 cm soil depth. Such weeds are classed as deep germinating weeds against the surface germinating (0.25 cm deep) and shallow germinating (up to 5 cm deep) ones. Therefore, a planting method that dries the top 3 to 5 cm of soil rapidly enough to deny weed seeds opportunity to absorb moisture for their successful germination, usually postpones weed emergence up untill the first irrigation. By this time the crop plants are well established to compete with the late germinating weeds. On the contrary, in a planting method that leaves the soil surface moist for considerable time after planting crops, the weed seeds absorb enough moisture to germinate readily. This induces severe weed-crop interference.

Crop Geometry

Crop density and rectangularity are very important in determining the quantum and quality of crop environment available for the growth of weeds. Wide row spacings with simultaneous high, intra-row crop plant populations may induce dense weed growth. Infact, in this situation there may develop additional, undesirable competition between the adjacent crop plants themselves. In this respect, square planting of crops in which there are equal row and plant spacings should be ideal in reducing within the crop plant competition. But in practice the density of crop plants seems to be more dominant than their rectangularity in deciding the course of weed-crop competition. For e.g.bi-directional sowing with recommended inputs or closer sowing (15 cm row to row spacing) have proved effective to provide more smothering effects on weeds as compared to wider (22.5 cm) row to row spacing in wheat crop. It is so because competition among individuals of the same species of plants is comparatively weaker than competition among individuals of the different species.

Crop Rotations

The association of a particular weed with a particular crop can be eliminated/ reduced by rotating the crops. For example *Phalaris minor* weed is associated with wheat crop and by replacing wheat with other crops like berseem, potato, winter maize, raya, gobhisarson etc., its intensity can be reduced or even eliminated. Similarly, *Echinochloa crusgalli* can be eliminated by replacing rice with other *kharif* crops.

Selection of Quick Growing Crops/Varieties

Weeds can be smothered very effectively by selecting quick growing crops or varieties within the same crop. Brassica crops have more smothering potential on Rabi weeds due to more leaf area than wheat. Similarly PBW 343 (aestivum varieties) has more smothering potential than durum varieties.

Increasing Plant Density

More number of plants of crop per unit area will cover the ground in a lesser time which helps in reducing weed menace. Dry matter of weeds in paddy can be reduced by increasing hills of rice from 33 (recommended) to 44 per sq.m.

Dab System

By providing two pre-sowing irrigations, the germination of weed seeds can be stimulated. These germinated weeds can be ploughed under or can be sprayed with contact herbicide. In this way seed reserves of field (seed bank) can be reduced. The other-Dab system in wheat is that field is left unsown for about 10-12 days after seed bed preparation and during this period repeated planking is done. The germinated seedlings of weeds are uprooted with planking and secondly the top 0-5 cm soil layer becomes dry which act as soil mulch. Wheat is sown slightly deep, so that its seeds are placed in moist soil zone. With the adoption of this practice weed population in the crop can be reduced.

Tillage

Tillage help in uprooting germinated seedlings while it also helps in bringing seeds from lower depths to upper soil surfaces. The seeds which have germinated with are-sowing irrigation can be killed with contact herbicides and crops can be sown without my tillage operation. ln this way also weed menace can be minimized.

Side Placement of Fertilizers

Placement of fertilizer especially nitrogen near to the seeding zone will provide a boost to the crop and weeds can be smoothened more effectively.

Intercropping

Growing of intercrop in the wider spaced crop will help in reducing the available space for the growth of weeds. So, weed problems can be reduced by growing inter-crops.

Use of Mulches

Mulches not only help in conserving soil moisture and modifying soil temperature but also cut down the germination rate of weeds. Due to non-availability of light, Reed seeds will not emerge where any type of mulch have been provided.

1.4

Weed Survey and Sampling Techniques

Weed monitoring is conducted to detect significant weed problems, rank them for relative severity and sort them by management type annuals, biennials, perennials, grasses and broadleaves. Weed monitoring provides information to tailor weed management efforts that reduce the impact of the weeds. The primary objective of weed survey and mapping is to accurately identify and delineate land with populations of unwanted plants. These surveys are conducted so that scientists and managers can predict those areas potentially subject to weed invasion; understand the biology of the invasion process and determine means by which weeds spread; develop, implement and evaluate weed management plans; assess the economic impact of weed invasion and increase public awareness, education and weed management efforts. Weed survey information is collected and compiled into maps showing the distribution and severity of infestation. Weed monitoring involves repetitive surveys to track weed populations over time. A standardized system of weed surveying and mapping is necessary to provide consistently reliable information that can be compared from year to year. Further, a standardized system allows weed survey data to be incorporated into a state wide weed survey database for the production of state wide noxious weed maps.

Objectives of Weed Survey

1. To collect information on weed biology and ecology, including location and acreage infested, growth requirements and spread patterns and rates.
2. Establishing a historical database.
3. Evaluation of the progress of a weed management programme.
4. Increase public awareness and support for weed management programme.

Methodology for Weed Survey/Surveillance

1. The agro-climatic zone in each state should be divided into sub zones and Route should be fixed for each zone for weed survey/surveillance.
2. Survey should be conducted twice in a year during both main seasons of kharif and rabi

3. Stoppage should be made at 10 km for each survey and the survey spot should not be near any building or permanent structure. One should walk at least 100 metre deep from main/link road in the ûeld for survey.
4. The minimum area under a particular crop should be one acre or more for Recording observations on Weeds and quadrate of 50cm X 50 cm or of any other size should be placed randomly at four spots in a ûeld moving diagonally.
5. At each ûeld the GPS data for longitude, latitude and height above mean sea level should be recorded.
6. Information from the farmer should be collected on following
   - Crop rotation for last three years.
   - Weed control methods used for last three years.
   - His experience with applied herbicide/weed control measure.
7. The observations on weeds should be recorded on:
   - Individual count of all weeds form each quadrate
   - New weed species infesting the area
8. Climate data should be taken from nearest weather station for the month in which survey is conducted and mean data for temperature, relative humidity, rainfall and sunshine hours should be reported. Ambient CO_2 concentration should also be recorded, if possible.
9. The each spot fixed through GPS should be surveyed again after 5 years to Monitor shift in weed flora, if any.
10. Work out the following values from the weed count data
    - Density/m^2= Total no. of individuals of sp. A in all the quadrates/total no of quadrates plotted.

(No. of individuals per 50cm X 50cm quadrate should be multiplied by 4 to obtain the density/m^2)

- Relative density (RD) (%)= (Density of sp. A x 100/sum density of all species)
- Frequency (%) = (No of quadrates where the species A occurred/ total no of quadrates plotted) x100

Relative frequency (RF) (%) = (Frequency of sp. A x 100/ Sum frequency of all species)

Importance value index (IVI) = Relative density + Relative frequency

The results of survey should be presented in the following way:

Route of survey: A——B——C——D (name of village or city)

Table: Weed flora of crop in the zone or district (——)

(GPS data for the range of latitude, longitude and height above MSL for the district)

Weed species (Total No.)	Density m^{-2}	Relative density (%)	Relative frequency (%)	Importance value index
Grass 1. 2. - - 10.				
Broad leaf 1. 2. - - 10.				
Sedges 1. 2. - - 10.				
Others				

(Only dominant weed species (about 10) should be listed and rest should be included in others which may be mentioned in text)

TNAU USDA-PL.480-AICRP on weed control-Survey

Proforma
Form A

Date of observation

1. Village :
2. PU/Taluk :
3. District /Region :
4. Name of the farmer :

	Wet	Garden	Dry
5. a) Total area of the farmer			
b) Name of the standing crop			
c) Name of the previous crops			
d) Tillage practices followed			
i) Preparatory cultivation			
ii) After cultivation			

6. Survey area details

a) Appearance of the site : Cultivable /fallow/forest/Waste land /roadside/pasture/marshy/canal/tank/river

b) soil :

i) Type : Red/Black/alluvial/laterite/Sandy/loamy/clay

ii) Texture :

iii) Soil reaction :

iv) Soil depth :

v) Calcareous or non-calcareous Acidic/alkaline/neutralShallow/medium/ deep

c) Source of manures and Fertilizers : FYM/Compost/sheep penning/Poultry or piggery GM. Or GLM and chemical fertilisers

d) Irrigation (i) Source : Well/tank/canal/river Brackish/moderate/good

ii) Quality :

iii) Frequency :

f) Any other information :

7. Signature of the observer :

Form B. Survey of Weeds

S.No.	Particulars	Location 1	Location 2	Remarks 3	Remarks 4
1.	**Sedges**				
	Cyperus rotuntus				
	C.difformis				
	C.iria				
	Fimbristylis miliacea				

2. **Grasses**
   Echinochloa crusgalli
   E. colonum
   Dactyloctenium aegyptium
   Digitaria sanguinalis
3. **Broad leaved weeds**
   Amaranthus viridis
   Ammania baccifera
   Portulaca oleracea
   Trianthema portucastrum
4. **New weed to be identified**

Important Points During Surveillance

1. The GPS data for each spot should be kept secured to be used at the time of Second cycle of surveillance which help in locating the spot
2. Any important information like invasion of some new weed, new emerging weed problem, shifts in. weed ûora, reduced efficacy of any herbicide against a particular weed(s) should be reported and highlighted in the text.
3. During interpretation of the results, Relative Density will tell about the comparative status of the population strength, Relative Frequency will tell about the status of distribution and Importance Value Index will tell about the overall dominance status of each species in the surveyed area.
4. Each year there should be 20% overlapping of the spots form the previous year.
5. Information on road side and waste lands should also be recorded.
6. Include five invasive weeds of the zone under survey in the surveillance programme.

Shift in Weed Flora

Weed management can be deûned as the manipulation of an agro ecological environment to create a situation favorable for crop growth but unfavorable for weed survival. Continuous application of a particular cropping practice can contribute inadvertently to a shift in dominance and distribution of weeds. In formulating a weed management program, it is important to systematically manipulate recommended production practices in synchronization with the current location-speciûc fanning activities.

Factors Affecting Composition of Weed Flora

Better understanding of the mechanisms of interaction between the different species in an Agricultural field, leads to selection of more effective weed management practices. The weed species found often will reflect the history of agricultural activity in the field. Broader understanding of the influence of chemical and nonchemical weed control treatments on weed population dynamics would enable us to predict ecological changes. This would facilitate adjusting treatments to prevent the weed population shifts that make control more difficult.

Important Factors Affecting the Species Composition are

- Soil tillage
- Water/irrigation
- Nutrient supply/fertilization
- pH/liming
- Crop rotation
- Planting method
- Planting season
- Seed dispersed at harvest
- Seed cleaning before planting
- Weed control methods
- Herbicide use pattern i.e. continuous/alternate use of herbicide group(s).

In Punjab, the introduction of rice has resulted in tremendous changes in weed flora. Before green revolution, rice was not the major crop of Punjab and *Eleusine aegyptiacum, Digitaria sanguinalis, Eragrostis tenella, Trianthema portulacastrum, Amaranthus viridis, Tribulus terrestris, Cenchrus catharticus, Digera arvensis, Euphorbia hirta* etc. were the major weeds of the kharif. Under such situations perennial weeds like *Cynodon dactylon, Sorghum halepense, Cyperus rotundus, Saccharum spontaneum* etc. were predominant during the kharif season. Change in cropping system particularly due to rice cultivation in Punjab has resulted into major shift in weed flora from typical summer season weeds to water lovingweeds like *Echinochloa crusgalli, Echinochloa colonum, Paspalum distichum, Cyperus iria* and with time new weeds like *Ischaemum rugosum, C. diformis, Fimbristylis* sp invaded therice ûelds. Further continuous cultivation of rice under flooded conditions changed the soil environment in favor of semi aquatic and few aquatic weeds such as *Leptochloa chinensis, Ammania basifera Caesulia axillaris, Sphenochlea zeylanica, Sagittaria guanensi, Ipomoea aquatica and Oxallis* sp. This showed that shift in weed flora was more rapid in the Punjabstate. Further due to canal irrigation system, typical weeds of hilly areas like *Ageratum conyzoides* invaded the rice field, water courses and bunds of the fields in the state. Development of semi aquatic conditions due to continuous cultivation of rice under flooded conditions encouraged species like *Scirpus roylei, Eleocharis* sp, *Najas arguta* the weed of marshy areas.

In wheat also, typical shift in weed flora of Punjab has taken place from *Carthamus oxycantha, Asphodelus tenuifolius* and *Avena ludoviciana to Phalaris minor, Poa annua, Polypogan monspeliensis, Malva neglecta* and *Rumex dentatus* with the development of better irrigation facilities. Recently it was observed *Coronopus didymus, Majus pumilus, Veronica agrestis* invade the wheat ûelds. Rice-wheat system lead to the dominance of Piularis minor where as non rice-wheat to *Avena ludoviciana, Rumex spinosis, Malvaneglecta* etc. Also zero till sowing of wheat created problem of broad leaf weeds especially Rumex dentatus.

An important contribution of rice cultivation in rice-wheat systems was the elimination of troublesome weeds viz. *Saccharum spontaneum, Saccharum munja, Cynodon dactylon, Sorghum halepense, Cyperus rotundus, Avena ludoviciana, Carthamus oxycantha, Asphodelus tenuifolius, Cirsium arvense, Convolvulus arvensis, Silene conidea* etc. The elimination of perennial weeds particularly of kharif season helped a lot in sowing of wheat with zero- till drill as after rice cultivation clean ûelds are available.

In cotton fields, survey of weed flora was conducted for twenty years, from 1985 to 2006, in the state of Punjab, six weeds viz. *Eleusine aegyptiacum, Trianthema portulacustrum, Digera arvensis, Cyperus rotundus, Digitaria ciliarus* and *Echinochloa colonum* remained predominant weeds of cotton. The weed survey/surveillance showed that some weeds namely *Cenchrus cathartichus* and *Tribulus terrestris* disappeared or are on the verge of disappearance and density of some weeds like *Eragrostis tenella, Sorghum halepense, Phyllanthus niruri, Physalis minima, Celosia argentia, Corchorus tridens* and *Cucumis calosus* reduced considerably during these years may be due to improved agronomic practices. Some new weeds like *Brachiaria reptans, Eragrostis diarrhena, Cyperuscompressus* etc. were found infesting cotton fields.

In maize crop, successful use of atrazine for weed control has resulted in invasion ofmaize fields by *Cyperus rotundus*. Continuous application of atrazine resulted in invasion ofmaize with *Brachiaria reptans, Acrachne racemose, Commelina benghalensis* etc. Theseweeds can be controlled with tank mix application of atrazine with pendimethalin, alachlor and trifluralin. Herbicides are used in orchards to eliminate broadleaved weeds and encourage agrass ground cover. Continues use of the same herbicide or herbicides that affect the same weeds, encourages unaffected genera because those susceptible to the herbicides are killed.

The periodic weed survey/surveillance showed that predominance of weeds keep on shifting form one weed to another weed mainly due to adoption of improved agronomic practices like change in cropping systems, better irrigation facilities, fertilizer application, sowing time, crop cultivars and particularly weed management/herbicide use. Therefore weed survey/surveillance in any crop gives an insight to the extent of weed problems and knowledge of associated weed composition which helps in devising weed management practices before it takes alarming proportions. So, regular surveillance should be made to formulate effective weed management programmes.

Absolute density (D) =

Total number of plants for a given species in all quadrats

$$\text{Relative density (Rd)} = \frac{\text{Absolute density for a given species (D)}}{\text{Total absolute density for all species}} \times 100$$

Absolute frequency (Af)

$$= \frac{\text{The number of quadrats in which a given species occurs}}{\text{Total number of quardrats (or) sample unit used (Max.10Nos.)}} \times 100$$

Relative frequency (Rf) =

$$\frac{\text{Absolute frequency value for a species (F)}}{\text{Total number of absolute frequency value for all species}} \times 100$$

Weed Sampling Techniques

The sampling techniques of weeds may be broadly grouped into (1) destructive and (2) non-destructive method. In the destructive method, the weed species are pulled out or fresh or dry weights are recorded. Dry weight of the weed species is more appropriate since the fresh weight may vary depending on the moisture content of the plant. In the non-destructive method, minimal damage is caused to the vegetation. In both the methods of sampling, quadrate is used as sampling unit. Three methods are used for weed sampling.

- Quadrate methods which method which include list or census quadrate and clip quadrate method.
- Points intercept method.
- Lines intercept method.

1. Quadrate Methods

The quadrate is a square area of varying sizes. Since the sample unit is a quadrate, it is popularly known as the quadrate method. In this method, sample area is marked off in the ûeld for the puropose of detailed weed study.

1. **List or Census Quadrate :** In this method, individual weed species are listed and the actual member of species per unit area is counted. It is easier to account the number of annual weed species having a single stem. But it is difûcult to count the species which propagate vegetatively. To overcome this problem, the rooted sprouted or individuals can be considered as individual number. While counting the species, care should be taken to count the species which fall within the quadrate only. The species which are lying or just over lopping the quadrate but not have from the sampling unit should be omitted. This method gives accurate assessment of abundance of a species. However, this method is tedious and time consuming. The accuracy depends on the size and the number of quadrates used.
2. **Clip Quadrate :** In this method each species in a quadrate is clipped off at the soil surface and the oven dry weight is recorded. Relative dry weight of a particular species can also be worked out.
3. **Point Intercept Method :** The method of sampling is followed in the pastures where the vegetation is dense or matted and difficult to determine where one individual species starts and another one ends. A sample unit is a frame having vertical row of slender pins equally spaced at 5 or O cm apart. When placed over the vegetation, only the stems of the species which have first touched by each pin point are recorded. Instead of a frame, a measuring length tape can also be used. It is stretched out for any desired length over the area to be sampled and is taken as the sample plot. The length should be constant for each sampled plot. Sampling pins are pushed vertically into the ground at any desired but constant interval along the tape. The weed species touched by the pins are recorded. If no species are contacted, that should also be noted.
4. **Line Intercept Method :** In dense stands of tufted plants where plant outlines are clear, this method of sampling can be used. In this method, a measuring tape is stretched out for any required length over the area to be sampled. This represents the sample plot. Length should be constant for each plot sampled. The intersect line is then subdivided into intervals of any desired length. The species which have touched the intersect line should be recorded.

Weed Vegetation Analysis

The weeds species are listed and number of individuals of each species is counted. This method is accurate, allows direct comparison of different areas and different species and is an absolute measure of the abundance of plant. The disadvantage of the method is that it I tedious and very time consuming. The following vegetation parameters are calculated.

Impotence value (IV) =

Relative density (RD) + Relative frequency (Rf)

$$\text{Summed dominance ratio (SDR)} = \frac{I.V.}{2}$$

In general, the density and population of weeds in experimental plots are commonly measured by either their weight or density or both. Weed weight by weed type and weed density by weed type both correlated well with crop yields. However combining weed weight and weed density did not appreciably improve correlations with yield over those based on either measure. The other derived parameters used for weed analysis are.

Summed Dominance Ratio (SDR)

The SDR again indicates the degree of dominance of a species over the other species in a given sample plot (Chul and Moody, 1983).

$$\text{SDR \%} = \frac{\text{IVI of the given species}}{\text{Number of parameters used to work out IVI}} \times 100$$

Weed Control Efficiency (WCE)

$$\text{WCI} = \frac{\text{Wc-Wt}}{\text{Wc}} \times 100$$

Where Wc is the average weed count or weight per unit area in the unweeded check plot, and Wt is similar information obtained from the treated plots.

$$\text{Weed Index (WI)} = \frac{Y_{WF}\text{-Yt}}{Y_{WF}} \times 100$$

Where Y_{WF} is the average crop yield in the weed free plot and Y_t is the average crop yield in the herbicide treatment under study.

$$\text{WCSE} = \frac{\text{Weed dry weight in control cum sole crop (kg/ha)} - \text{Weed dry weightin weed management cum smothering intercrop (kg/ha)}}{\text{Weed dry weight in control cum sole crop (kg/ha)}}$$

In experiments dealing with the determination of the efficacy of some inter-crop on the suppression of weeds vis-a-vis sole cropping, the WCE in the first formula is replaced by. Weed Smothering Efficiency (WSE) which is given as:

$$WSE = \frac{Mdw\text{-}ldw}{Mdw} \times 100$$

Where Mdw and ldw are the average dry weight obtained from the sole main crop plots and the intercropped plots, respectively.

Weed Control Cum Smothering Efficiency (%)

Weed Community

Another part of vegetative study deals with the techniques to compare the areas of different weed communities, relate vegetation variation to environment factors and to identify the controlling factors. The parameter used for these aspects are :

Co-Efficient of Community (C)

Two communities or the vegetation stands of two regions can be compared by using mathematical expression of the similarities of the lists of species. If community X is compared to Y, the number of species common to both accounted and expressed as a percentage to the total number of X plus Y (Pablico and Moody, 1983).

Co-Efficient of Similarity (C)

By using quantitative data such as I.V. and SDR for various species, pairs of communities may be compared by calculating co-efficient of similarities (C).

$$C = \frac{2W}{a+b} \times 100$$

Where C = co-efficient of similarity; W = some of the lower I.V. or SDR of all species of first community and be sum of I.V. or SDR of all species of second community.

The values (C) are calculated based on the absolute and relative values of vegetation parameters. They may not be identical but are comparable. The community similarity co-efficient can be converted to a dissimilarity co-efûcient (D) by the equation.

$$D = 100—C$$

Simpson's Index (S)

This index is used to measure the concentration of dominance and determine the degree of diversity in a given community type.

$$S = \sum \left\{ \frac{Y}{N} \right\}^2$$

Where, Y = I.V. of a given species

N = sum of I.V. for all species in the sample (100)

Weed Sample Collection, Preservation and Preparation of Herbarium

Plants are considered as weeds when they interfere with human activity in this endeavour of cultivating crops of economic nature. In order to manage them effectively one should know their identity, character and behaviour in the crop environment. Morphological heterogeneity occurs within same species often leads to task of identifying them difficult. Collection and preservation of weeds in a particular environment in the form of herbaria is the first step as these forms a ready reference for the future as well help in effective management. A collection of plants in preserved form is technically known as herbarium.

Materials, Methods and Procedures

Collection of Weeds

- Collect weeds which are common in the field environment
- Care should be taken to collect them intact with all parts i.e. roots, stem, leaves, flowers, fruits and seeds.
- Two to three specimens may be collected per species
- It should be representative of the species
- Select plants, which are free from insect pest and disease damage.
- Large plant should be represented by selected portions such as branches, reproductive structures and any special features
- Ideal size of the specimen is about 40 cm
- Avoid collecting samples when they are moist
- Avoid plants which are withered
- The best time for collection is during afternoon hours of sunny days
- Assign a collection number for each species sequentially and never to be repeated

Observation to be Recorded During Collection

- Place and date of collection with collectors name
- Colour of the flower, leaf, stem and fruit
- Individual part wise hairiness on the entire plant
- Rooting pattern and distribution
- Growth habit and habitat
- Odour of leaves and flowers
- Record special features if any

Pressing the Specimen

- Remove soil and mud by washing with the fresh water and then wipe with dry cloth
- Place plants carefully in a folded newspaper
- Arrange in a way so that important features of the vegetative and reproductive parts are clearly seen including lower and upper surface of leaves
- If specimen is too long it should be bent in to a V or U or N shape
- Avoid overlapping and crumbling
- Seeds that are shed readily should be placed into paper envelopes and placed with the specimens.
- Pile the specimens on top of each other, each being separated by absorbent boards.
- Press the pile gently between a pair of wooden presses (apply only light pressure)
- Attach an identifying tag with number already assigned

Drying the Specimen

- The specimens are allowed to dry under shade in a well-aerated place
- Care should be taken that the specimens are not infected with fungal moulds
- Change the moisture absorbent boards depending up on the specimens moisture content
- Observe every day or every other day and avoid over drying
- Sun drying retains the natural colour
- Artificial drying can be done in the temperature ranging from 45 to 60 °C

Use of Preservatives

- Weed specimens once dried should be chemically treated to protect them from insect attack and from other destructive organisms as herbariums in tropical countries are frequently prone to termite attack.
- Dipping the specimens in any one of the poisoning solution is suggested
- Twenty-five grams mercuric chloride and 25 grams cresol dissolved in 1 litre industrial alcohol.
- Fifteen grams of mercuric chloride and 10 grams phenol crystals dissolved in 1 litre denatured alcohol. or
- A 20 per cent solution of lauryl pentachlorophenate dissolved in methylated spirits.
- Proper care should be taken in handling these solutions as they are corrosive and re dry the plants.

Mounting the Specimens

- Use standard size heavy sheets of 42x29 cm for mounting the dried plants
- Mount the specimens on sheets using various fixing materials like cellophane tapes, suitable adhesives etc.
- Attach herbarium label containing collection number, scientific name, family, Collector's name and date of collection.

Storage

- Mounted specimens may be stored in closed containers
- Place cloth bag containing 1:1 mixture of Para dichlorobenzene and naphthalene and replenish at periodical intervals
- The storage room may be sprayed with mild insecticide regularly.

Unit-2: Weeds Control

2.1

Weed Control Methods

Methods of Weed Control

Prevention, eradication and control are the fundamental principles of weed control.First basic step is to prevent the infestation of a weed(s) in a locality in order to avoid their further spread. If one fails in preventing spread of weeds, eradication step must be adopted. Eradication is limited on a very small scale and adopted only when suitable methods of complete weed removal are available as eradication is cent per cent control of weeds alongwith its root system and other propagative parts. Further, if one fails in prevention and eradication techniques then different weed control methods must be strictly followed in which the objective is to bring down the weed population (with any method) to a level that economical crop yields are possible. The aim of each weed control method is to achieve reasonable good control of weeds. Following methods for controlling weeds are very common:

Preventive Methods

Try to prevent spread of weed seeds/vegetative parts with the adoption of following preventive techniques.

Use of Clean Seeds

Care should be taken at the time of sowing of crops that weed seed free crop seedmust be used. Before sowing crop seeds, thorough sieving should be done in order to sieve out weed seeds, broken and shrivelled grains and diseased seeds. Check the seeds thoroughly before sowing and ensure that no weed seed is present. If it is not possible to separate out the weed seeds due to their similar size to that of crop seed, better reject that lot for seed purposes.

Use of Well Rotten Farm Yard Manure

Farm yard manure (FYM) is a big source of weed seeds because weeds are usuallyfedalong with fodder to the animals. Weed seeds in the viable form are excreted out of the digestive track along with excreta (cow dung). Generally farmers do not decompose FYM properly and apply in raw/undecomposed form in the field and such ûelds develops higherweed infestations. Farmyard manure, is no doubt, very valuable organic manure but should be used in well rotten form. Generally farmers store FYM in open heaps where proper temperature required for rotting (to kill weed seeds) are not achieved. So, avoid the use of raw FYM and always use well decomposed FYM to get rid of addition of weed seeds into the weed seed bank of soil as well as infestation of termites.

Uprooting Weeds Before Flowering

When weed population is scanty and it is possible to uproot these plants, remove these before flowering and seed setting. This preventive technique will help in reducing/eliminating weed seed bank, which will be beneficial for the future years. In case any stranger/new Weed species is seen in the field it should be uprooted before flowering in order to prevent future infestation. This technique also helps in managing herbicide resistance in weeds.

Keep Clean Field Boundaries and Irrigation Channels

Perennial weeds multiply in the permanent bunds and then enter in the field. So, before sowing a crop, clean the field boundaries physically or spray non residual herbicides like glyphosate. Similarly irrigation channels should be kept free from weeds in order to improve water use efficiency as well as to reduce weed infestations. The weed plants growing on the banks of irrigation channels/canals should not be allowed to set seeds as these seeds disperse with flowing water and infest new areas.

Use of Clean Farm Machinery

All the types of machinery must be cleaned while entering into the new field. All thetype of cultivating equipment must be cleaned alter ploughing weedy fields in order to remove weed seeds (sticking to soil) and vegetative parts of perennial weeds. Proper care must be taken while moving cultivating implements/ harvesting combines from one field to another field. Lot of weed seeds remains in different chambers/parts of combines/farm implements and there is a need to clean each and every component of implements/combines especially after operation. Running combines without load for few minutes help in cleaning different components with air.

Nursery Samplings

Many new weed species disperse from one to another place with the transport of nursery of fruit/ornamental/vegetable plants. While sampling these plants, proper care should betaken that all off type plant should be removed. After transplanting these samplings in the field, remain vigilant and remove all new weed species, if any, before flowering particularly from the close vicinity of the fruit plant.

Cultural Methods

Cultural weed control includes non-chemical crop management practices ranging from variety selection to land preparation to harvest and post harvest processing. It includes tillage, planting, cultivars, plant population, time of fertilizer application, irrigation. Cultural weed control is a part of integrated weed management which involves the integrated use of cultural, manual, and/or mechanical control methods.

Crop Rotation

Crop rotation is the best and non-monetary technique for weed management because weeds are associated with certain crops due to their identical ecological requirements. The problem of *Trianthema portulacastrum* can be controlled by cultivating puddled rice. Similarly *P. minor* population can be eliminated/ reduced by replacing wheat with berseem, potato, raya/gobhisarson, etc.

Growing of Intercrops

Inter cropping suppresses weeds better than sole cropping and thus provides an opportunity to utilize crops themselves as tools of weed management. Many short duration pulses *viz.,*green gram and soybean effectively smother weeds without causing reduction in the yield of main crop.

Stale Seed Bed

A stale seedbed is one where initial one or two flushes of weeds are destroyed before planting of a crop. This is achieved by soaking a well prepared field with either irrigation or rain and allowing the weeds to germinate.At this stage a shallow tillage or non- residual herbicide like paraquat may be used to destroy the dense flush of young weed seedlings. This technique allows the crop to germinate in almost weed-free environment.

Blind Tillage

The tillage of the soil after sowing a crop before the crop plants emerge is known as blind tillage. It is extensively employed to minimise weed intensity in drill sowing crops where emergence of crop seedling is hindered by soil crust formed on receipt of rain or irrigation immediately after sowing.

Date of Sowing

After studying biology of a troublesome weed, the sowing of the crop can be manipulated in such a way that ecological conditions for germination of weed seeds are not met due to escape mechanism. It has been seen that first flush of *P. minor* which is hard to control can be avoided by enhancing the sowing of wheat crop during the recommended period (last week of October) when the prevailing temperature is higher which does not suit for *Phalaris minor* seeds to germinate (Virk *et al.*, 2003).

Plant Density

The objective of increasing plant density is to provide minimum space to weeds to grow due to overcrowding of crop plants, resulting thereby increased competitive potential by the crop. Heavy stand of crop will cover ground in a shorter period of time, thereby cutting/reducing penetration of light on the ground where weeds are present. Bhan, 1987 reported significant decrease in dry matter accumulation by weeds by increasing seed rate of wheat from 100 kg to 150 kg/ha.

Planting Pattern/Crop Srchitecture

The objective of this technique is uniform distribution of plants of crop per unit area even with same seed rate by providing better crop architect. Mahajan*et al.*, 2000 reported significant decrease in dry matter accumulation by *P. minor* and significant increase in grainyield of wheat by reducing row to row spacing from 22.5 cm to 15.0 cm.

The seed rate is being constant. So, proper (close) plant to plant spacings results in poor growth of *P. minor* as indicated by significant reduction in dry matter of *P. minor.*

Method of Fertilizer Application

Placement of fertilizer is very beneûcial in providing lead to the crop due to quick and more availability of nutrients to crop plants as compared to weed plants. Ahuja and Yaduraju (1989) reported significantly less *P. minor* plants/m^2 in placement method (2.5 cm below seed) as compared to broadcast method of fertilizer application.

Selection of Quick Growing Varieties

Varieties with initial quick growth and more leaf area must be preferred in order to reduce crop weed competitions. Virk *et al.* (2003) compared different wheat varieties for smothering *Phalaris minor*. They reported that bread wheat cultivars (PBW 154 and PBW 343) reduced dry matter accumulation by *Phalaris minor* significantly than durum wheat cultivar i.e. PDW 233 and hence resulted in significant increase in grain yield.

Tillage

Primary role of tillage is to up root the germinating or germinated weed seedlings. In conservative tillage system, weed seeds present in the top soil zone are forced to germinate with pre-sowing irrigation and can be killed with any contact herbicide and crops can be sown without any tillage operations. With the adoption of this method weed population in crop is expected to come down.

Summer Tillage

The practice of summer tillage or off-season tillage is one of the effective cultural methods to check the growth of perennial weed population in crop cultivation. Initial tillage before cropping should encourage clod formation. These clods, which have the weed propagules, upon drying desiccate the same.

Dab System

It is a cultural technique to allow the weed seeds present in the soil to germinate before sowing of the crop and sufficient time is provided to weed seeds to germinate before sowing the crop. Either double pre sowing irrigation is given or field is kept as such for 8-10 days (during winters only) after land preparation to encourage weed seeds to germinate. These germinated weeds can be ploughed up at the time of land preparation or can be killed with the use of any contact type herbicide.

Straw mulch

Polythene mulch

Mulching

Mulch is a protective covering of material maintained on soil surface. Mulching has smothering effect on weed control by excluding light from the photosynthetic portions of a plant and thus inhibiting the top growth. It is very effective against annual weeds and some perennial weeds like *Cynodon dactylon.* Mulching is done by dry or green crop residues, plastic sheet or polythene film. To be effective, the mulch should be thick enough to prevent light transmission and eliminate photosynthesis.

Suppression of weeds with the use of mulch may be attributed to various chemical physical factors. Physical factors of mulch respond for reduced weed seed germination due to lower soil temperature and light penetration. Weed population and dry weight of weeds were greatly reduced in maize crop when mulched with wheat straw (Crutch field *et al.*, 1982). Mulching is more useful in wider row crops as compared to narrow row crops.

Irrigation Management

Irrigation management has direct or indirect effect on weed intensity. In drip irrigation the weed intensity is generally lower than in furrow irrigation method. Similarly, in alternate furrow irrigation, low weed intensity was observed in dry furrows than in irrigated furrows.

Soil Solarisation

Soil solarisation is a novel technique of controlling soil born pests including weeds. This is another method of utilisation of solar energy for the desiccation of weeds. It involves covering the wet soil with thin transparent polythene/ polyethylene films during the summer months. The process would raise the surface soil temperature by 8-12°C as compared to non-solarized soils. The plastic sheet checks the long wave back radiation from the soil and prevents loss of energy by hindering moisture evaporation. A duration of 4 to 6 weeks is sufficient to give satisfactory control of weeds. Many annuals, some perennials and parasites weeds are sensitive to this treatment. Besides controlling soil born pests, soil solarisation is reported to enhance availability of nutrients in soil and further benefit of microflora, ultimately resulting in increased plant growth response in many crops.

Eradiation

Eradication is complete elimination of all live weeds, weed parts and weed seeds. It can be defined as 100% and complete control.

Eradication means elimination of weeds after they have become established in an area and control means methods which are adopted where prevention and eradication have failed,so control measure as a rule eradicate the weed and make it possible to raise the crops in spite of their presence.

1. Ideal weed control but rarely achieved
2. Very difficult & high cost
3. Can be attempted in small areas/high value areas
4. Green/Polyhouses (Protected Cultivation), Ornamental Plant Beds, Containers

Mechanical Methods

Mechanical or physical methods of weed control are being employed ever since man began to grow crops. Removing the weeds physically or with tools/ implements is called mechanical method. The mechanical methods include tillage, hoeing, hand weeding, digging cheeling, sickling, mowing, burning, flooding, mulching etc. Different mechanical methods adopted for controlling weeds are as follows:

Tillage

Tillage is done mainly with two objectives. First is to uproot the weeds growing in thefield and second to pulverize the soil in order to create ideal conditions for the seeds to germinate. There are different implements which can be used for this purpose such as cultivators, disc harrows, mould board plough, deep chisler etc. Tillage removes weeds from the soil resulting in their death. It may weaken plants through injury of root and stem pruning, reducing their competitiveness or regenerative capacity. Tillage also buries weeds. Tillage operation includes ploughing, discing, harrowing and leveling which is used to promote the germination of weeds through soil turnover and exposure of seeds to sunlight, which can be destroyed effectively later. In case of perennials, both top and underground growth is injured and destroyed by tillage. Deep tillage helps in reducing soil seed bank to a great extent by burying the weed seeds to lower depths. However, these days few contact and non-residual herbicides are available and with the use of these chemicals emerged weeds can be controlled and the frequency of tillage can be reduced.

Hand Weeding

It is done by physical removal or pulling out of weeds by hand or removal by implements called khurpi, which resembles sickle. It is probably the oldest method of controlling weeds and it is still a practical and efficient method of eliminating weeds in cropped and non-cropped lands. It is very effective against annuals, biennials and controls only upper portions of perennials. This method is laborious and cumbersome but common for controlling weeds from transplanted crop of rice and from nurseries. However, this method should be encouraged in all field crops to uproot scanty (few) plants of weeds, which are left with the adoption of any weed control method. In order to reduce soil seed bank, this method should be practiced before flowering and seed setting stage of the weeds.

Hoeing

Hoe has been the most appropriate and widely used weeding tool for centuries. It is however, still a very useful implement to obtain results effectively and cheaply. It supplements the cultivator in row crops. Asmall implement like khurpa (khurpi), kasolla, spade, baguri, wheel hoe etc. can be used for uprooting weed plants from different crops. In all crops, the time of hoeing is very critical andit should neither be done too early nor too late but crop should be kept free from weeds atcritical period of crop weed competition. Hoeing is particularly more effective on annuals and biennials as weed growth can be completely destroyed. In case of perennials, it destroyed the top growth with little effect on underground plant parts resulting in re-growth. This method is very safe but the only limitation is non-availability of labour when required and is very expensive method. Hand hoeing can bedone in all row-sown crops which may be solid drill or widely sown.

Interculture

This method is only possible in wider row sown crops such as maize, cotton, sugarcane, pigeon pea etc. Interculture can be done with bullocks or with tractor by adjusting the distance between the tines so that crop rows are not hampered. It is very quick method ascompared to hand hoeing and this method can be practiced over a large area. Moreover, this method is very effective for controlling latter flushes of weeds. In cotton, sugarcane and maize, this method is very widely adopted. The limitations of this method is that weeds growing in the vicinity of crop plants or in the intra row areas are not controlled and sometimes the crops which have lateral spread of roots suffer due to root injury. Even the spread of plant diseases may also takes place in few crops such as potato.

Digging

Digging is very useful in the case of perennial weeds to remove the underground propagating parts of weeds from the deeper layer of the soil.

Sickling

Sickling is also done by hand with the help of sickle to remove the top growth of weeds to prevent seed production and to starve the underground parts. It is popular in sloppy areas where only the tall weed growth is sickled leaving the root system to hold the soil in place to prevent soil erosion.

Burning

Burning or fire is often an economical and practical means of controlling weeds. It is used to (a) dispose of vegetation (b) destroy dry tops of weeds that have matured (c) kill green weed growth in situations where cultivations and other common methods are impracticable.

Flooding

Flooding is successful against weed species sensitive to longer periods of submergence in water. It is very effective method for controlling annual weeds requiring aerobic conditions and many perennials such as *Cyperus rorundus*, *Cynodon dactylon*, *Sorghum halepense* etc. According to *Gill and Walia (1987)* continuous flooding in puddled rice helpsin reducing weed infestations. This method helps killing weeds due to exclusion of oxygen availability with puddling operation. The dry matter of weeds was significantly less when duration of ponding water was for 9 days and hence significant increase in seed yield was obtained as compared to the treatment where ponding of water was maintained for 3 days only. The success of flooding depends upon complete submergence of weeds for longer periods.

Flaming and Steaming

Weeds are desiccated with high-pressure stream or even with flames in orchards and plantations. This method can also be implemented in wider row sown crops. Flaming and steaming method was practiced in many countries, especially in plantation crops.

Mowing

Mowing is a machine-operated practice mostly done on roadsides and in lawns.Mowing is very effective method for cutting excessive growth of undesirable plants from lawns, playgrounds, orchards, roadsides and from uncultivated areas. This method is effective for annual weed control; however, perennials resprout from root stock. However, inperennial weeds due to repeated cuttings, the food reserve of the root stock is exhausted anddue to non-availability of required food, weed may die due to starvation. This method apart from controlling weeds also enriches the soil by adding organic matter.

Merits of Mechanical Method

1. Oldest, effective and economical method
2. Large area can be covered in shorter time
3. Safe method for environment
4. Does not involve any skill
5. Weeding is possible in between plants
6. Deep rooted weeds can be controlled effectively

Demerits of Mechanical Method

1. Labour consuming
2. Possibility of damaging crop
3. Requires ideal and optimum specific condition

Mechanical Weeders

Dry Land Weeder

It is used for weeding in row crops for removing shallow rooted weeds. It has been designed ergonomically for easy operation. Useful in dryland and gardenland crops and is ideal at a soil moisture content of 8 to 10 per cent.

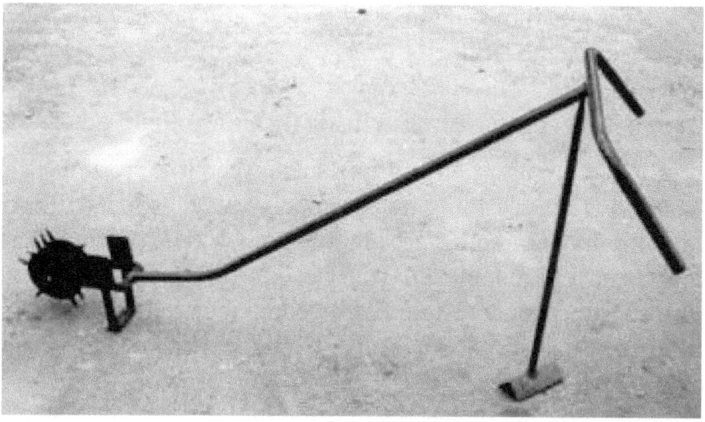

At the extreme end of the arm 120 mm diameter star wheel is fixed. A cutting blade is fitted to the arm 200mm to the back of the star wheel the star wheel facilitates easy movement of the tool. The operating width of the blade is 120 mm. Ideal to remove shallow rooted weeds. The workable moisture content has to be 8 to 10 %

Power Rotary Weeder

For mechanical control of weeds in crops such as sugarcane, tapioca, cotton, tomato and pulses whose rows spacing is more than 45 cm.

The rotary weeder consists of three rows of discs mounted with 6 numbers of curved blades in opposite directions alternatively in each disc. These blades when rotating enable cutting and mulching the soil. The width of coverage of the rotary tiller is 500 mm and the depth of operation can be adjusted to weed and mulch the soil in the cropped field.

Tractor Drawn Weeding Cum Earthing up Equipment

For weeding and intercultural operations in between row crops in a single pass an inter cultivator cum earthing up equipment was developed and fitted to a standard tractor drawn ridger. Three number of sweep type blades are affixed to the ridger frame for accomplishing the weeding operation in between standing rows of crops. Three ridgerbottom fitted behind the sweep blade, work on the loosened soil mass and aid in earthing up by forming ridges and furrows. Weeding efficiency is 61 per cent.

Tractor Operated Multi Row Rotary Weeder

For weeding and intercultural operations in between row crops like sugarcane, cotton, maize, etc. in a single pass.

The multi row rotary weeder consists of a set of cutting blades, which penetrate in to the soil, removing the weeds in the crop rows. The cutting blade has also been used as an inclined plane for elevating and converging the soil. The rotating blades are used to cut the weeds and pulverizing the soil. Weeding efficiency is 71 per cent.

Chemical Method

Control of weeds with the use of chemicals called herbicides that are toxic to certain categories of plants is called as chemical weed control method. The idea regarding control ofweeds with herbicides started when Bordeaux mixture which was sprayed to cure a particular disease of wheat in Europe and it showed killing of broadleaf weeds. Similar types of reportsregarding control of broadleaf weeds with copper sulphate were published by Bolley (1908) from USA, Schulty (1909) from Germany and Bonnett (I909) from France. Similarly Robate(1917) in France conducted experiments with copper sulphate, ferrous sulphate and sulphuricacid to control weeds. Later on efficacy of more inorganic salts such as sodium arsenate,sodium chlorate, potassium chlorate, sodium borate, potassium borate etc. was reported. During 1941, Pokorny in USA reported synthesis technique of 2,4-D. Zimmemam & Hitchcock from USA during 1942 reported use of 2,4-D as growth regulators. Hammer andTukey (USA) during 1944 reported that at higher concentrations 2,4-D can act as herbicide.Nutman and Blackman from England confirmed these ûndings during 1945. So, the history of weed science dates back to the use of 2,4-D on commercial scale for the control of broad leaf weeds from cereals. Later on, the introduction of triazine, substituted ureas, carbamates,dinitroanilids, sulfonylureas and others new molecules brought a revolution in herbicide useagainst weeds in different crops.

Advantages of Chemical Method

As compared to mechanical method, chemical weed control method has many advantages, which are discussed as follows:

Eliminate Early Crop Weed Competition

In order to obtain higher yields, weeds must be controlled before the critical period ofcrop weed competition that lies generally between 2-6 weeks after sowing of most of the crops. This period varies with crop and season in which it is raised. Pre-emergence as well as post-emergence application of herbicides are done before critical period of crop-weed competition starts. So, chemical method of weed control provides good environment for the crop for its initial growth and development.

Morphologically Similar Weeds to Crops can be Effectively Controlled

Few weeds particularly during initial stages have similar morphological appearance as of crop. For example, *Echinoch-loacrusgalli* resembles with rice (in nursery) and *Phalarisminor* as well as *Avenaludoviciana* with wheat at the time of hoeing. So, identification of these weeds is a big problem and these plants are left in the field considering them as crop plants. The seedlings of *E.crusgalli* are oftenly uprooted with rice nursery and are transplanted in the field along with rice nursery. These transplanted plants of *Echinochloacrusgalli*are not killed with any herbicide and these produce huge quantity of seeds for the next years. So, with the use of selective herbicides in rice nursery and in wheat, these morphologically similar weeds can be controlled very effectively.

Intra Row Weeds can be Controlled Effectively

Intra row weeds are not controlled with hand hoeing/interculture and only inter row weeds are removed with this practice. But on the other hand both inter and intra-row weeds can be effectively controlled with the use of herbicide due to their selectivity phenomenon. So, the uncontrolled intra row weeds in mechanical method compete with crop for growth factors and also produce seeds which act as source of infestation during the succeeding years.

Sure Method of Controlling Weeds

Sometimes due to continuous rains or due to non-availability of timely labour, mechanical method cannot be implemented. With delayed hoeing, not only labour expenses are increased but overall net returns are reduced due to severe competition with weeds. However, there is less risk of rainfall or other climatic factors on the adoption of chemical method.

Holds Good in Problematic Soils

The ûelds, which are very clayey in texture, do not come in proper moisture condition for hand hoeing and these operations are either delayed or omitted due to continuous wet field conditions. Moreover, hoeing of such fields results in re-establishment of uprooted weeds. On the other hand, with the use of herbicides weeds can be effectively controlled from these type of situations.

Problematic Weeds can be Effectively Controlled

Few weeds such as *Trianthema portulacastrum* and *Phalaris minor* have a tendency to germinate in flushes and such Weeds are not controlled effectively with mechanical means.Herbicides with ideal residual periods provide good control of such category of weeds.

Herbicides Kill Weeds *in situ*

The weed plants are killed *in situ* (where they are) with the use of herbicides. However, with mechanical methods uprooted weeds are dragged to new site where they may re-establish if soil is clayey in texture and optimum moisture is present. This problem is comparatively more serious during winter months or in *Rabi* crops.

No Injury to Root

Pruning of the roots of crops takes place with hoeing or interculture technique which may prove harmful for overall crop development. However, such type of losses do not occur in chemical weed control method.

Tillage can be Minimized

Frequency of tillage operations depends upon weed density along with type of weedflora. Crops can be sown without any tillage operation with the use of non-residual also called zero-till herbicides. So, use of herbicides is substitute of tillage operations which cutdown the cost of cultivation to a great extent. The use of glyphosate or paraquat can be successfully made depending upon weed flora to minimize tillage operations. Prefer glyphosate where perennial weeds are present.

Economical Method

The daily wages of farm labour are increasing day by day and thereby cost of two hand weeding practice comes out to be very high. On the other hand per acre cost of chemical weeding is much low as compared to mechanical weeding.

Limitations of Chemical Method

Chemical weed control method, though played key role in green revolution but simultaneously it had created new problems in today's agriculture. Most important limitations are.

Lack in Application Technology

In general farmers do not adopt proper application technology for spraying herbicide. The spray techniques are faulty and majority of the farmers move spray lance to and for which is absolutely wrong practice. Always spray in bands (strips) by fitting either fat fan food jet nozzles. Do not use cone type nozzle for spraying herbicides as its discharge is lowand its distribution pattern is not uniform. Mostly farmers do not stick to the recommended dose and time of application. Due to these drawbacks desirable control of weeds is no attained.

Weed Resurgence/Resistance

Continuous use of one group of herbicide may lead to either resurgence or development of resistance in weeds. The problem of resurgence is very common in perennial weeds evenwith the adoption of best weed control methods which may be due to their strong root system which contain enough food reserves. For instance resurgence has been noticed in *Cyperus rotundus* and *Sorghum halepense* with hand hoeing or with the use of glyphosate herbicide. On the other hand, the continuous use of isoproturon for the control of *Phalaris minor* in wheat grown in north-eastern states of India particularly Punjab and Haryana, had created problem of resistance in *Phalaris minor* to isoproturon which may be due to increased metabolism of this herbicide by the weed plants. So, in order to avoid such problems, rotational use of different herbicide groups should be made so that all types of weeds may be controlled.

Shift in Weed Flora

The applied herbicides do not kill all weed types (species) very effectively or at the same level which are present in the field. Few weed plants are less susceptible (slightly tolerant) to the applied herbicide. So, continuous use of same group of herbicide may result in shift inweed flora. For example, with the regular use of butachlor, the infestation of *Ischaemumrugosome* and broad leaf weeds are on the increase in transplanted crop of paddy whereas population of *Echinochloa crusgalli* have decreased.

Herbicide Drift

Drift or run-off of herbicide can cause damage to the neighbouring crops. When high velocity winds are blowing at the time of spraying, some sensitive crops growing in adjoining fields may be damaged. For example, while spraying 2, 4-D in sugarcane field, its drift may do much harm to the adjoining cotton fields. Under high temperature conditions, volatilization of applied herbicide may harm sensitive crops even to a long distance. Similarly with heavy rain fall, run off losses of applied herbicides can damage sensitive crops.

These can be Used as Insecticides by Mistake

The left-off stock of pesticides (insecticides, fungicides and herbicides) is mostly stored on one self by the farmer without any proper labelling. By mistake farmer can use herbicideas insecticide and vice versa. Sometimes, farmer had to bear great loss of crop due to this mistake. So, proper labelling and storing of herbicides should be ensured.

When Crop Fails, no Choice for Succeeding Crops

Sometimes crop do not germinate either due to poor seed quality or due to toxic effect of herbicide. Under such situations farmer is left with no option. For example GSL-2, variety of gobhisarson is resistant to atrazine and when this herbicide is sprayed by mistake on anyother variety of gobhisarson, the crop fails. During Rabi season, there are very meagre chances of crop(s) which can tolerate residues of atrazine. Similarly if wheat crop of the farmer treated with sulfosulfuron fails due to excessive rains or any other reason, the farmer is left with no or very limited options for selection of crop during *Rabi* season which can tolerate to the residual effect of this herbicide.

Chemical Crop War

Some mischievous farmers use these chemicals as weapon for taking revenge of theirenemies (neighbourers). During odd hours they spray or broadcast non-selective herbicides on their fields (of enemies) and slowly crop is totally damaged.

Cost of Few Herbicides are High

Few herbicides due to their high cost are out of the reach of an average farmer. There is no doubt that net benefit is increased to a higher level with the use of such herbicides. However, only disadvantage is that their initial cost is high.

Pollution Hazards

When herbicides are used in excess their residues/persistence can be detected in crop produce, soil and even in underground water. These residues are harmful for human and animal health. Apart from this the long soil persisting herbicides prove hazardous for succeeding crops. So, an ideal pesticides must be eco friendly with desirable residual effects. Continuous use of herbicides over a longer period may even pollute underground water Radix, *et al.* (1994) analyzed the drinking water for pesticide contamination in an intensive agricultural region in the south east of France which showed alachlor, atrazine, chlorotoluron, mecoprop and simazine at significant levels (100-1000 mg/lit) in the drinking water.

No Antidote Available

If unluckily crop shows herbicide toxicity symptoms due to over dose or with the use ofwrong (unrecommended) herbicide, no antidote (safener) is available in the market which canovercome herbicide toxicity symptoms. Under such type of situations, farmer had no alternative but to loose the crop.

Biological Methods

Biological suppression is an effective, environmentally safe, technically appropriate, economically viable and socially acceptable method of pest management. Biological method has gained too much popularity for the control of insect pests followed by fungal diseases in agricultural crops. Scope of biological control of weeds is limited in the country due to many reasons. Firstly due to akin morphology of crop and weeds, the selectivity basis is limited. Secondly lot of weeds (different species) infest crop and bioagent are specific to one species only.

The biological control of weeds involves the use of living organisms, such as insects, disease organisms, herbivorous fish, other animals and competitive plants to limit their infestations.

Qualities of Bioagents

1. **Host specific :** Bioagents must be very specific to host and they should not attack other economic plants growing nearby. The insects selected for biological control should prefer to starve to death rather than feed upon plant species other than target weed.
2. **Adjustment to field environment :** After raring the insects in the lab, their test for survival are done under field conditions. So, an ideal bioagent is that which can survive in the ûeld without any loss in its activity.
3. **Easy to multiply:** High rate and ease of natural reproduction of bio agents is important in insects, pathogens, snails and competitive plants. This quality of bioagent is important to limit (reduce) the multiplication time.
4. **Feeding habits : More** stress should be laid on the selection of insects as bioagents that attack either the flower or seeds of weed plants or bore into the stem. Likewise, insects which attack roots are likely to be very effective on perennial weeds. The destructive capacity of a bio agent on the weed should be very high sothat its small population could destroy large weed infestations.

Kinds of Bio agents

Four types of bio agents have been identified which can be successfully used for controlling terrestrial and aquatic weeds.

1. **Insects:** Insects have been more frequently used than other animals for the biological control of weeds because these are very host speciûc i.e. one insect species is employed to destroy only one weed species.
2. **Plant pathogens:** Many fungi are observed to attack specific weed species, but examples of successful practical weed control with fungi are few. However, this technique has very wide scope for the control of weeds in crops and on non-cropped area with the use of mycoherbicides.

Bio-herbicides

The use of plant pathogens for killing the target weeds, which are made available in spray able formulation like chemical herbicides by artificial culturing, is termed as bio-herbicide. Fungal pathogens of weed have been used to a larger extent than bacterial, viral or nematode pathogens, because bacteria and virus are unable to actively penetrate the host and require natural opening or vectors to initiate disease in plants.

Mycoherbicide

A fungal pathogen, which when applied inundatively, kill plants by causing a disease is called as mycoherbicide. Here the specific fungal spores or their fermentation product is sprayed against the target weed. Some registered mycoherbicides in western countries are tabulated below.

S.No.	Product	Content	Target weed
1.	Devine	A liquid suspension of fungal spores of *Phytophthora palmivora* causes root rot.	Strangle vine (*Morrenia odorata*) in citrus
2.	Collego	Wettable powder containing fungal spores of *Colletotrichum gloeosporoides* causes stem and leaf blight	Joint vetch (*Aeschyomone virginica*) in rice, soybean
3.	Bipolaris	A suspension of fungal spores of *Bipolaris sorghicola*	Jhonson grass (*Sorghum halepense*)
4.	Biolophos	A microbial toxin produced as fermentation product of *Steptomyces hygroscopicus*	Non-specific, general vegetation

3. **Carp-fish :** Certain fresh water carp-fish consumes large quantities of aquatic weeds and they are very suitable for relieving water bodies of their massive growth of noxious, aquatic vegetation. Chinese grass carp is the most promising species of carp for aquatic weed control. This carp-fish eats weeds more than its body weight daily.

4. **Competitive plants:** Certain plants which are highly competitive in nature are capable of suppressing some specific weeds without causing any meaningfulhinderance to our interests. *Panicumpurpurascens*Raddi is found competitive with *Typha* spp. in marshy lands. Similarly *Cassia cerassia* is very competitive planteliminating *Partheniumhy-sterophorus* (Congress grass), from non-cropped area.

Few Examples of Biological Control of Weeds are as Follows

1. ***Lantana camara*** The flowers of *Lantana camara* are orange, white, light blue in colour. Seeds are small preferred by birds and dispersal through their droppings. Million of acres were infested in Hawaii Island up to 1900. Few insects were imported from Argentina and Mexico. From these *Plusiaverticillata* provided successful control of this weed.

2. ***Opuntiainermis* (Prickly pear)** During 1900 about 10 m acres were infested with this weed in Australia which rose to 60m acres by 1925. Twelve insects were imported from Argentina, USA and Mexico. Most effective insect (moth) was *Cactoblastic-cactorum*and about 90% area was cleared upto 1930.

3. ***Cyperusrotundus*** *Bactra verutana* a shoot boring moth was reported to control this weed in USA, India and Pakistan.
4. ***Parthenium hysterophorus*** The beetle namely *Zygogramma-bicolorata* provided biological control of this weed in India during rainy season by feeding its flowers and leaves.
5. ***Ludiwigia parviflora*** is completely denuded by *Altica cynanea* (steel blue beetle).

6. **Annual and perennial weeds** Weeds can be controlled by grazing animals from the crops which are pungent intaste.Goats and sheep can be successfully used for controlling weeds from Japanese mint (*Mentha arvensis* and other species) when the crop is due for hand hoeing. These animals only graze the weeds and no harm is done to the crop. These animals provide very effective control especially annual weeds because due to their typical grazing habits, annual weeds are uprooted at the time of grazing. Pigs are very effective for controlling perennial weeds *Sorghum halepense, Cyperus rotundus, Cynodon dactylon* etc. because these animals are able to dig out underground vegetative reproductive parts of these weeds.

7. **Control of aquatic weeds** Several attempts have been made on the biological control of these weeds, both by pathogens and with insects. Most success in this respect has been met in Florida (USA) with a Hyacinth moth i.e., *Sameodesal biguttalis* B. Benner which is native of South America (Center, 1982). Its larval feeds upon young leaves and apical buds of waterhyacinth. *Salvinia molesta* (Salvinia) a water fern with bioagent namely *Cytrobagous salviniae* Sands. Alligatorweed in Florida (USA) can be effectively controlled with flea beetle (*Agasicleshygro-phylaselman* Vogt.) and several bayous have been cleared of this weed there. Herbivorous fish Tilapia controls algae. Common carp, a non-herbivorous fish controls sub-mersed aquatic weeds. It is apparently due to uprooting of plants while in search of food. Snails prefer submersed weeds.

Limitations

This method could not gain its popularity due to following limitations :

1. After feeding on the host plant, these insects may move to economical plants. So, there ismore risk of damage to economical plants.
2. Raring of the insects is a big problem. For raring insects on large scale in the laboratory, lot of space as well as different types of substrate are required.
3. Biological control is very costly method of weed control because screening of insects on large scale is done in lab and majority of insects do not adjust to environmental conditions. Later on, when released in the field only few can surviveto the prevailing conditions. So, lot of investment is required for screening desirable predators.
4. Isolation of desired area which we want to keep-off the insect is not possible as there is no available method with which we can achieve this target.

Biotechnological Methods of Weed Control

Success of Herbicide Resistance Crops (HRC) in Indian and World Agriculture

Excessive weed growth forces the crops to compete for sunlight & nutrients leading to yield losses & herbicides do not differentiate between plants. Conventional agricultural use "selective herbicides" wherein herbicides do not harm the crop but are not effective at removing all types of weeds. In Herbicide resistance crops (HRC) 'non-selective' herbicides can be used to remove all weeds in a single & quick application. Less spraying & lower operating costs.

Herbicide Tolerant Crops can be produced either by insertion of foreign gene from another organism into the crop or by regenerating herbicide tolerant mutants from existing crop germplasm. First process produces Genetically Modified Organism (GMO) and the second process produce non-GMO variety or hybrid. GMO herbicide tolerant crops include canola, soybean varieties, corn hybrids tolerant to glyphosate and gulfosinate herbicides. Non-GMO herbicide tolerant crops include sulfonylurea-tolerant soyabean, clear field corn and wheat. 2012 marks the 17th anniversary of the commercialization, 1996-2012, of biotech crops (genetically modified (GM) or transgenic crops). The experience of the first 16 years of commercialization, 1996 to 2011, has confirmed that the early promise of crop biotechnology has been fulfilled. Biotech crops have delivered substantial agronomic, environmental, economic, health and social benefits to farmers and, increasingly, to society at large. The rapid adoption of biotech crops reflect the substantial multiple benefits realized by both large and small farmers in industrial and developing countries, which have grown biotech crops commercially. Between 1996 and 2011, developing and industrial countries contributed to a record 80-fold increase in the global area of biotech crops from 1.7 million hectares in 1996 to 160 million hectares in 2011.

There is a growing body of consistent evidence across years, countries, crops and traits generated by public sector institutions that clearly demonstrate the benefits from biotech crops. These benefits include improved weed and insect pest control with biotech herbicide tolerant and insect resistant Bt crops, that also benefit from lower input and production costs; biotech crops also offer substantial economic advantages to farmers compared with corresponding conventional crops. The severity of weeds, insect pests and diseases varies from year-to-year and country to country, and hence will directly impact pest control costs and the economic advantages of biotech crops in any given time or place.

Table 2.1.2: Global Status of Commercialized Biotech/GM Crops 2012 (Top 10 countries)

Country	Area under GM Crop (m ha)	Crops
USA	69.5	Maize, Soybean, Cotton, Canola, Sugarbeet, Alfalfa
Canada	11.6	Maize, Soybean, Canola, Sugarbeet
India	10.8	Cotton
Brazil	36.6	Maize, Soybean, Cotton
Argentina	23.9	Maize, Soybean, Cotton
China	4.0	Cotton, Papaya, Tomato,Sweet Pepper,
Paraguay	3.4	Maize, Soybean, Cotton
South Africa	2.9	Maize, Soybean, Cotton
Pakistan	2.8	Cotton
Uruguay	1.4	Soybean, maize

Dominant Biotech Crop in 2012

Herbicide tolerant soybean continued to be the dominant biotech crop in 2012, occupying 80.7 million hectares or 47% of global biotech area (Table 3). It was grown commercially in the USA, Argentina, Brazil, Paraguay, Canada, Uruguay, Bolivia, South Africa, Mexico, Chile, and Costa Rica.The second most dominant crop was biotech maize with stacked traits, which occupied 39.9 million hectares or 23% of the global biotech area. It was grown commercially in 10 countries including the USA, Brazil, Argentina, South Africa, and Canada, the Philippines, Uruguay, Honduras, Chile and Paraguay.Bt cotton was the third most dominant crop grown in 2012. Bt cotton was planted in more than 18.8 million hectares in 13 countries: India, China, Pakistan, Myanmar, Burkina Faso, Brazil, USA, Argentina, Australia, Colombia, Sudan, Paraguay, and Costa Rica. This is equivalent to 11% of the global biotech area.

Table 3: Dominant Biotech Crops in 2012

Crop	Million Hectares	% Biotech
Herbicide tolerant soybean	80.7	47
Stacked traits maize	39.9	23
Bt cotton	18.8	11
Herbicide tolerant canola	9.2	5
Herbicide tolerant maize	7.8	5
Bt maize	7.5	4
Stacked traits cotton	3.7	2
Herbicide tolerant cotton	1.8	1
Herbicide tolerant sugar beet	0.5	<1
Herbicide tolerant alfalfa	0.4	<1
Others	<0.1	<1
Total	170.3	100

Source: Clive James, 2012

Global Adoption of Biotech Soybean, Maize, Cotton, and Canola

Another way to provide a global perspective of the status of biotech crops is to characterize the global adoption rates as a percentage of the respective global areas of the four principal crops-soybean, cotton, maize and canola, in which biotechnology is utilized.In 2012, 81% (80.7 million hectares) of the 100 million hectares of the soybean planted globally were biotech (Figure 3). Biotech cotton was planted to 24.3 million hectares, which is 81% of the 30 million hectares of global cotton. Of the 159 million hectares of global maize planted in 2012, 35% or 55.1 million hectares were biotech maize. Finally, herbicide tolerant biotech canola was planted in 9.2 million hectares or 30% of the 31 million hectares of canola grown globally in 2012. If the global areas (conventional and biotech) of these four crops are aggregated, the total area is 320 million hectares, of which 53% or 170.3 million hectares were biotech, up from 50% in 2011. Two-thirds of these 320 million hectares are in developing countries farmed mainly by small, resource-poor farmers.

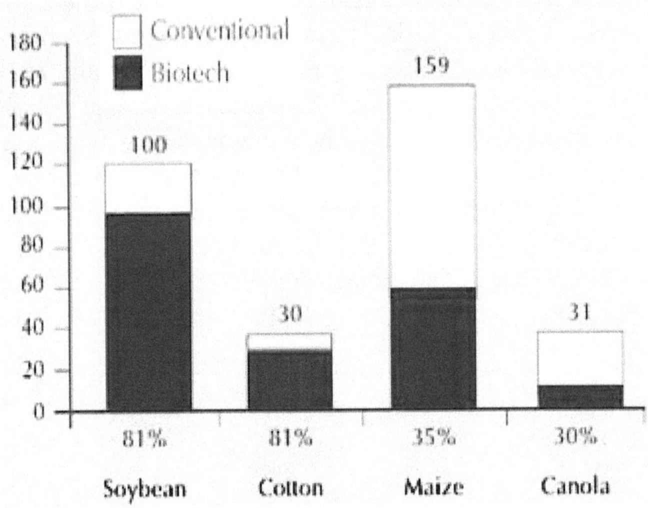

Fig. 3: Biotech Crop Area as % Global Area of Principal Crops, 2012 (Million Hectares). *Source:* Clive James, 2012

Benefits of Herbicide Tolerant Crops

- Broad spectrum of weed control
- Reduced crop injury
- Reduced herbicides carry over
- Use of herbicides that are more environmental friendly
- New mode of action for resistance management
- Crop management flexibility and simplicity

Concerns with Herbicide Tolerant Crops

- Yield performance
- Single selection pressure and weed resistance
- Shift in weed species
- Gene escape
- Gene flow and contamination of organic crops
- Drift and non - target movement

Future Prospects

The first stacked soybean (with herbicide tolerance and insect resistance traits) will be planted in Brazil in 2013; vitamin-A enriched Golden Rice could be released in the Philippines in 2013 or 2014; drought tolerant sugarcane in Indonesia; and biotech maize in China. Biotech crops is not a panacea; but they have the potential to make a substantial contribution to the 2015 MDG goal of cutting poverty by half, by optimizing crop productivity, which can be achieved by public-private sector partnerships.

Integrated Method

For a long time, herbicides were mainly seen as a substitute to cultural and mechanical methods of weed control. The need for developing an integrated approach of weed controlwas already clearly recognised about few years ago. In the meantime, however, the situation for the control of weeds has seriously deteriorated due to drastic and wide spread emergence weeds that are resistant to currently used herbicides. To date, more than 70 undesired plant species have been found to be resistant to all triazine compounds. Further, with the continuous application of the same chemical herbicide and routinely repeated cultivation practices has led to modification in weed communities. Social and environmental concern, as well as a desire to improve weed control efficiency has led to increase emphasis on integrating several cultural methods, although herbicides, screening trials still is in force inconventional weed research in India. A model on integrated weed control given by Noda, 1977 is as shown in Fig 1.

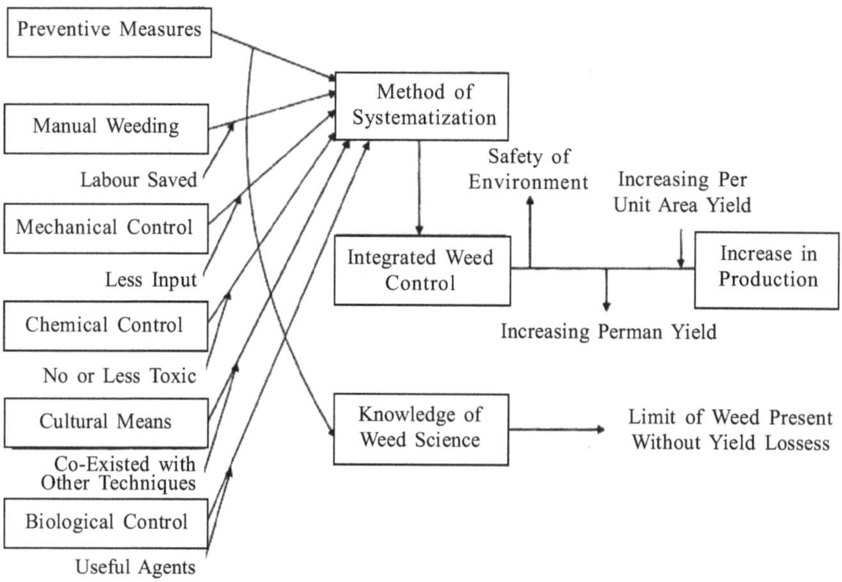

Fig. 1: A conceptional model of integrated weed control

Integrated Weed Management (IWM) is a method whereby all economically, ecologically and toxicologically justifiable methods are employed to keep the harmful organisms below the threshold level of economic damage. Keeping in the foreground the conscious employment of natural limiting factors (FAO definition), it can also be deûned asthe rational use of direct and indirect control methods to provide cost-effective weed control.

Among the commonly suggested indirect methods are land preparations, water management plantspacings, seed rates, cultivar use, fertilizer applications etc. Direct methods include cultural, mechanical and chemical methods of weed control. The essential factor inany IWM programme is the number of indirect and direct methods that can be combined economically in a given situation.

2.2

Integrated Weed Management

Integrated Method of Weed Management

Definition

i) An integrated weed management may be defined as the combination of two or more weed-control methods at low input levels to reduce weed competition in a given cropping system below the economical threshold level. It has proved to be a valuable concept in a few cases, though much is still to be done to extend it to the small farmers' level.

ii) When more than one method is employed for controlling weeds, it is termed as integrated method weed management.

FAO Definition

It is a method whereby all economically, ecologically and toxicologically justifiable methods are employed to keep the harmful organisms below the threshold level of economic damage, keeping in the foreground the conscious employment of natural limiting factors.

IWM is the rational use of direct and indirect control methods to provide cost-effective weed control. Such an approach is the most attractive alternative from agronomic, economic and ecological point of view.

Integrated Weed Management Approach

Integrated Weed Management (IWM) approach aims at minimizing the residue problem in plant, soil, air and water. An IWM involves the utilization of a combination of mechanical, chemical and cultural practices of weed management in a planned sequence, so designed as not to affect the ecosystem. The nature and intensity of the species to be controlled, the sequence of crops that are raised in the rotation, the standard of crop husbandry, and the ready and timely availability of any method and the economics of different weed-management techniques are some of the potent considerations that determine the success for the exploitation of the IWM approach.

This method is more effective because the left over weeds with one method can be controlled with other method. So, this method helps in reducing seed bank status in the field. With the adoption of this method, many problems such as shift in weed flora, development of resistance in weed plants etc. can be avoided. Integrated method of weed control is the need of the day and this method must be advocated in order to get long term relief from these undesirable plants. Integration of chemical, mechanical, cultural or even biological method can be made for effective and long time weed control. Integrated weed management approach is environmentally friendly as farmers are not entirely dependent on herbicides. So, need based use of herbicides should be done and if need be it can be integrated with other weed control methods.

Concept

- Uses a variety of technologies in a single weed management with the objective to produce optimum crop yield at a minimum cost taking in to consideration ecological and socio-economic constraints under a given agro-ecosystem.
- A system in which two or more methods are used to control a weed. These methods may include cultural practices, natural enemies and selective herbicides.

Good IWM should be

a) Flexible enough to incorporate innovations and practical experiences of local farmers.

b) Developed for the whole farm and not for just one or two fields and hence it should be extended to irrigation channels, road sides and other non-crop surroundings on the farm from where most weeds find their way in to the crop fields.

c) Economically viable and practically feasible.

Advantages of IWM Include

- Shifts in crop-weed competition in favour of crops,
- Prevents weed shifts towards perennial nature,
- Prevents weed resistance to herbicides,
- Danger of herbicide residues in soil or plant can be minimized,
- Minimum environmental pollution, and
- Contribute to economic crop production.

Integrated Weed Management in Agricultural Crops

Cereals

Rice

Nursery

i) Annual grasses and sedges can be controlled with the application of 1.5 kg/ha of butachlor or thiobencarb 7 days after sowing of pre germinated (sprouted) rice seeds after mixing in sand. These herbicides can also be applied on a moist soil 3 to 7 days before puddling and sowing of pre germinated rice seed. Alternatively application of Sofit 30.5 EC (Pretilachlor + Safener) at 1.25 lit/ha (=0.38 kg a.i./ha) as sand mix, 3 days after sowing/broadcasting of pre-germinated rice seeds. A post-emergence herbicide i.e. bispyribac sodium can also be used for controlling weeds in rice nursery.

(OR)

ii) Apply any one of the Pre-emergence herbicides viz., Butachlor 2 l/ha, Thiobencarb 2/ha, Pendimethalin 2.5 l/ha, Anilofos 1.25 l/ha on 8th day after sowing to control weeds in the lowland nursery. Keep a thin film of water and allow it to disappear. Avoid drainage of water. This will control germinating weeds.

Transplanted Rice

Pre-emergence

i) Use Butachlor 2.5 l/ha or Thiobencarb 2.5 l/ha or Fluchoralin 2 l/ha or Pendimethalin 3 l/ha or Anilofos 1.25 l/ha as pre-emergence application.

ii) Alternatively, pre-emergence application of herbicide mixture viz., Butachlor 1.2 l + 2,4-DEE 1.5 l/ha or Thiobencarb 1.2 l + 2,4-DEE 1.5 l/ha or Fluchoralin 1.0 l + 2,4-DEE 1.5 l/ha or Pendimethalin 1.5 l + 2,4-DEE 1.5 l/ha or Anilofos + 2,4-DEE ready mix at 1.25 l/ha followed by one hand weeding on 30-35 days after transplanting will have a broad spectrum of weed control in transplanted rice.

(OR)

Pre-emergence application of pretilachor at 1.0 kg a.i. ha^{-1} on 3 DAT + weeding with Twin row rotary weeder at 40 DAT

(OR)

PE Pyrazosulfuron ethyl 10 % WP @ 150 g ha-1 on 3 DAT + hand weeding (HW) on 45 DAT.

(OR)

PE butachlor 0.75 kg a.i. ha^{-1} + bensulfuron methyl 50 g ha on 3 DAT + HW on 45 DAT

(OR)

PE Oxadiazon 87.5 g a.i. ha^{-1} followed by Post emergence (POE) 2,4-D 1kg a.i. ha along with hand weeding on 35 DAT.

(OR)

PE butachlor 0.75 kg per hectare + bensulfuron methyl 50 g ha^{-1} on 3 DAT followed by mechanical weeding on 45 DAT is effective for broad spectrum weed control.

iii) The weeds in rice can be controlled with the use of herbicides listed below within 3 days of transplanting after mixing with 50 kg of sand or fertilizers and applied uniformly to the field keeping 2.5 cm depth of water. These herbicides are very effective against annual grasses. In order to get desired control of weeds minimum gap should be kept between puddling-transplanting-application of herbicides. Efficiency of the applied herbicide will improve with continuous stagnation of water for at least one week. Moreover, neither irrigation nor drainage is given for two days from the date of application of herbicide.

Herbicides recommended for transplanted paddy as pre-emergence application

Name of herbicide	Dose (kg/ ha)
Butachlor 50 EC	1.5
Thiobencarb 50 EC	1.5
Anilofos 30 EC	0.375
Pendimethalin 30	0.75 to 0.90
ECPretilachlor 50	0.75
ECPyrazosulfuron-ethyl	0.015
10 WPOxadiargyl 80 WP	0.09

Early Post-emergence

i) Early post emergence application of Bispyripac sodium 50 g a.i. ha^{-1} (2-3 leaf stage of weeds) + Hand weeding on 45 DAT.

Post-emergence

i) All the traditional weeds of rice i.e. grassy, BLW and sedges can be controlled with the post-emergence (25-30 days after transplanting-DAT) application of bispyribac sodium 10 SC (Nominee Gold/ Wash Out/Macho/ Taarak). This herbicide is applied through spray application at 25 g per ha with Knap sack sprayer after draining the excess water from the field.

However, if rice seedlings are transplanted on beds then apply bispyribac sodium at 30 g per ha. Fields can be flooded one to two days after spray. For controlling broad leaf weeds and sedges from transplanted rice, many herbicides are very effective. For controlling these weeds, post-emergence application (20-25 DAT or when the weeds are in 3-4 leaf stage) of metsulfuron (Algrip Royal) at 15 g ha^{-1} (or) 2,4-D sodium salt (Fernoxone 80% WP) 1250 g ha^{-1} can be done through spray application after draining out excess water from the field. Alternatively, ethoxy sulfuron (Sunrice 15 WDG) 20 g ha^{-1} bensulfuron ethyl (Londex 60 DF) 60 g ha^{-1} or azim sulfuron (Segment 15DF) 20 g ha^{-1} can be sprayed 15-20 DAT after draining excess water from the field.

Cultural Method

i) Hand pull the germinated weeds twice or thrice on 15th and 30th DAT in order to eliminate their competition.

ii) Continuous ponding for at least 3-4 weeks also helps a lot in reducing/ controlling nearly all weeds of transplanted paddy.

iii) Dual cropping of rice-azolla, and rice-green manure (described in wet seeded rice section 2.5 & 2.6 of this chapter) reduces the weed infestation to a greater extent.

iv) Summer ploughing and cultivation of irrigated dry crops during post-rainy periods reduces the weed infestation.

Late Hand Weeding

i) Hand weed a second time, 80-85 days after transplanting, if necessary.

Mechanical Method

i) Use of rotary weeder from 15 DAT at 10 days interval. It saves labour for weeding, aerates the soil and root zone, prolongs the root activity, and improves the grain filling though efficient translocation and ultimately the grain yield.

Wet Seeded Puddled Lowland Rice

i) Pre-emergence application of pretilachlor 0.75 kg/ha on 8 DAS or pretilachlor + safener (Sofit) at 0.45 kg/ha on 3-4 DAS followed by one hand weeding on 40 DAS in direct drum seeded rice

ii) Sowing with drum seeder and Cono weeding (manual / power weeder) is done at 10, 20 and 30 DAS

iii) Productivity and economic returns of wet seeded rice with dual cropping of danicha could be maximized by the pre-emergence application of pretilachlor + safner at 0.45 kg ha^{-1} followed by one cono weeder in between rows and manual weeding with the rice rows on 35 DAS in lowland conditions.

iv) PE pretilachlor + safner 0.45 kg/ha^{-1} on 3 DAS + roto cylindrical weeder weeding on 45 DAS in wet seeded rice resulted in excellent control of weeds. PE pretilachlor (S) 0.45 kg ha^{-1} on 3 DAS fb azimsulfuron 50 DF 35 g ha^{-1} on 20 DAS + hand weeding on 45 DAS for broad spectrum weed control and higher grain yield and economic returns in both irrigated and rainfed direct seeded rice.

v) Higher productivity of wet direct seeded (drum seeded) rice could be achieved by integrating intercropping of daincha and pre-emergence application of pretilachlor + safner at 0.45 kgha^{-1} on 4 DAS followed by one hand weeding on 35 DAS.

vi) In rice-rice -fallow system intercropping of *Sesbania rostrata* control the weeds of rice field along with incorporation of *Sesbania rostrata* in to the field and one hand weeding on 35DAT.

vii) Apply PE pretilachlor 0.45 kg ha^{-1} on 3 DAS + Roto cylindrical weeds + weeding on 45 DASin wet seeded rice have good control of weeds like *Echinochloa crusgalli, Panicum repens,Eclipta alba* and *Monochoria vaginalis*.

Rainfed Rice (Dry Seeded Rainfed Un-Puddled Lowland Rice)

i) First weeding can be done between 15 and 21 days after germination.Second weeding may be done 30-45 days after first weeding.

ii) Apply pendimethalin 1.0 kg/ha on 5 days after sowing or Pretilachlor + safener (Sofit) 0.45 kg/ha on the day of receipt of soaking rain followed by one hand weeding on 30 to 35 days after sowing.

iii) Seed drill sowing with pre-emergence application of pretilachlor + safener @ 0.3 kg/ha followed by two weedings with star / rotary weeder is recommended.

Semi Dry Rice (Dry Seeded Rainfed un-puddled Lowland Rice with Supplemental Irrigation)

i) First weeding can be done between 15 and 21 days after germination. Second weeding may be done 30 - 45 days after first weeding.

ii) Apply pendimethalin 1.0 kg/ha on 5 days after sowing or Pretilachlor + safener (Sofit) 0.45 kg/ha on the day of receipt of soaking rain followed by one hand weeding on 30 to 35 days after sowing.

iii) PE butachlor 1.0 kg ha^{-1} followed by weeding using finger type single row and double row rotary weeders resulted in higher grain yield and net profit.

Semi Dry Rice (Dry Seeded Irrigated Un-Puddled Lowland Rice)

i) First weeding should be done between 15 and 21 days after germination.

ii) Second weeding may be done 30-45 days after first weeding.

iii) Apply pendimethalin 1.0 kg/ha on 5 days after sowing or Pretilachlor + safener (Sofit) 0.45 kg/ha on the day of receipt of soaking rain followed by one hand weeding on 30 to 35 days after sowing.

SRI-Rice (System of Rice Intensification)

i) Using rotary weeder/Cono weeder/power operated two row weeder, moving the weeder with forward and backward motion to burry the weeds and as well as to aerate the soil at 7-10 days interval from 10-15 days after planting on either direction of the row and column. Manual weeding is also essential to remove the weeds closer to rice root zone.

Aerobic Rice

i) Pre emergence application of pendimethalin at 0.75 kg/ha followed by two hand weeding or mechanical weeding on 25 and 45 DAS

ii) PE pendimethalin 1.0 kg ha^{-1} along with single tyne sweep weeding on 45 DAS which was comparable with PE along with hand weeding.

Wheat

i) Spray Isoproturon 800 g/ha as pre-emergence spraying 3 days after sowing followed by on hand weeding on 35th day after sowing.

ii) If herbicide is not applied, give two hand weeding on 20th and 35th day after sowing.

iii) Closer sowing (15 cm), high seed rate (120-150 kg/ha) or bidirectional sowing with isoproturon (0.5) as pre-emergence, skip row sowing (45-15-15-15-15-45 cm), hoeing and rouging of weeds, application of isoproturon (0.94), inducing germination by early irrigation or killing weeds by preparatory cultivation, using paraquat, closer sowing and high seed rate. In *Phalaris minor* dominated ecosystem higher seed rate (150 kg/ha) with square planting (20 x 20 cm) and application of isoproturon (1.0) is recommended.

Maize

The weeds can be controlled either mechanically or chemically.

i) Two hand hoeing 2 and 4 weeks after sowing keeps the weeds under check. Alternatively, application of atrazine (Atrataf 50 WP) as pre-emergence at 1.0 kg/ha in 400-500 litres of water provides weed free environment to this crop over a longer period. This herbicide take care of all annual grass and broad leaf weeds of maize. Atrazine can be sprayed as early post-emergence i.e. within 10-15 days after sowing at the same rate. However, its band application (20 cm wide strip on crop rows) can be made by using l/3 of recommended herbicide. The weeds from inter row area can be controlled with some contact herbicide like Gramoxone or with intercultural operations. The fields which are free from the infestation of itsit/chupati (*Trianthema portulacastrum*) can be treated with alachlor (Lasso 50 EC) at the rate of 2.5 kg/ha as pre-emergence. This herbicide is very effective against annual grasses. The continuous application of atrazine in the maize growing belt of Punjab resulted in partial development of resistance in few weeds namely Bans patta (*Brachiaria reptans*), Kaon makki (*Commelina benghalensis*) and Gha (*Acrachne racemose*) etc. For the control of these hardy weeds, tank mix application of atrazine 0.75 kg ha^{-1} with pendimethalin 0.75 kg ha^{-1}/ alachlor 1.25 kg ha^{-1}/trifluralin 1.20 kg ha^{-1} were found very effective. These herbicides also provide good control of other annual weeds of maize. For the control of Dila/Motha (*Cyperus rotundus*), apply 2, 4-D amine salt 58% at 0.58 kg a.i/ha as post emergence 20-25 days after sowing maize crop by dissolving in 400 lit/ha of water. For cultural control of weeds grow one or two rows of fodder cowpea in between two maize rows and harvest cowpea 35-45 days after sowing before it starts twining the maize plants. No weed management is required to maize as all the grown up weeds are harvested with cowpea fodder crop. For maize + Soybean intercropping system, apply pre-emergence Alachlor at 4.0 l/ha or Pendimethalin 3.3 l/ha on 3rd day after sowing as spray.

ii) Hoe and Hand weed on 17th or 18th day of sowing if herbicide is not applied.

Note: If pulse crop is to be raised as intercrop, do not use Atrazine.

Millets

Sorghum

i) Apply the pre-emergence herbicide Atrazine 50% WP 500 g/ha on 3 days after sowing as spray on the soil surface, using Backpack/knapsack/Rocker sprayer fitted with a flat fan nozzle using 900 lit of water/ha

ii) Sorghum is slow growing in early stages and is adversely affected by weed competition. Therefore keep the field free of weeds upto 45days. For this, after pre-emergence herbicide application, one hand weeding on 30-35 days after sowing may be given.

iii) If pulse crop is to be raised as an intercrop in sorghum do not use Atrazine.

iv) Hoe and hand weed on the 10th day of transplanting if herbicides are not used. Hoe and weed between 30-35 days after transplanting and between 35-40 days for direct sown crop, if necessary.

Ratoon Sorghum

i) Remove the weeds immediately after harvest of the main crop

ii) Hoe and weed twice on 15th and 30th day after cutting.

Rainfed Sorghum

i) Keep sorghum field free of weeds from second week after germination till 5th week. If sufficient moisture is available spray Atrazine @ 500 g/ha as pre-emergence application within 3 days after the soaking rainfall for sole sorghum while for sorghum based inter-cropping system with pulses, use Pendimethalin 3 l/ha.

Cumbu

Transplanted Crop

i) Spray Atrazine 50 WP 500 g/ha on the 3rd day. Then, one hand weeding on 30-35 days after transplanting may be given. If herbicide is not used hand weed on 15th day and again between 30-35 days after transplanting.

Direct Sown Crop

i) Apply the pre-emergence herbicide Atrazine at 500 g/ha, 3 days after sowing as spray on the soil surface using Back-pack/Knapsack/Rocker sprayer fitted with flat type nozzle using 900 lit of water/ha.

ii) Apply herbicide when there is sufficient moisture in the soil.

iii) Hand weed on 30-35 days after sowing if pre-emergence herbicide isapplied.

iv) If pre-emergence herbicide is not applied hand weed twice on 15 and 30 days after sowing.

Ragi

i) Apply Butachlor 2.5 l/ha or Fluchloralin 2 l/ha or Pendimethalin 2.5 l/ha, using Back-pack/Knapsack/Rocker sprayer fitted with flat fan type of nozzle with 900 lit of water/ha.

ii) Apply herbicide when there is sufficient moisture in the soil or irrigate immediately after the application of herbicide.

iii) If pre-emergence herbicide is not applied hand weed twice on 10th and 20th day after transplanting.

iv) For rainfed direct seeded crop, apply post emergence herbicide; 2,4-DEE or 2,4-D Na salt at 0.5kg/ha on 10th day after sowing depending on the moisture availability.

Pulses

Blackgram, Greengram, Soybean, Redgram, Bengalgram, Field Pea, Cowpea and Lentil

Pulse crops are having good smothering potential against weeds and during latter phases of crop growth all these crops can smother second or third flush of weeds.

i) Mechanical method: Weeds from pulses can be kept under control with 1-2 hoeings with khurpi/hand hoe, or wheel hoe etc. particularly during initial stages of crop growth.

ii) Cultural method: If herbicide / hoeing is not given, give two hand weeding on 15 and 35 days after sowing.

iii) Chemical fb HW/hoeing: Spray Fluchloralin 1.5 l/ha or Pendimethalin 2 l/ha 3 days after sowing mixed with 900 l of water using Back-pack/Knapsack/Rocker sprayer fitted with flat fan type nozzle. Then irrigate the field. Following this one hand weeding may be given 30-35 days after sowing **(For Blackgram, Greengram, Redgram, Bengalgram and Cowpea).**

Weeds can also be controlled with the use of selective herbicides fb hand hoeing/HW which are listed as follows (Table):

Herbicides recommended for pulses

Name of the crop	*Selective herbicides*	Dose (kg a.i ha^{-1})	Time of application
Moong (Greengram)	Trifluralin (Treflan 48 EC)	0.96	Pre- plant
	Pendimethalin 1.0 kg ha^{-1*}+	*	Pre-emergence
	Imazethapyr @ 1.0 kg ha^{-1*}+HW on 30DAS	*	Pre-emergence
Mash	Pendimethalin (Stomp 30 EC)	0.75	Pre-emergence
	Pendimethalin + one HW	0.45	Pre-emergence
Soybean (Irrigated)	Pendimethalin (Stomp 30 EC)+1HW on 30DAS	0.45	Pre-emergence
	Fluchloralin (Basalin 45 EC)+1HW on 30DAS	0.90	Pre-emergence
	Alachlor (Losso 50 EC)	2.00	Pre-emergence
Soybean (Rainfed)	Fluchloralin (Basalin 45 EC)	0.90	Pre-emergence
	Imazythypur +1HW on 30 DAS.	0.40	Pre-emergence
	Alachlor (Losso 50 EC)	2.00	Pre-emergence
Gram (Bengalgram)	Trifluralin (Treflan 48 EC)	1.20	Pre- plant
Field Pea	Pendimethalin (Stomp 30 C)	0.75	Pre-emergent:
	Pendimethalin (Stomp 30 C) + one HW	0.45	Pre-emergence
Lentil	Trifluralin (Treflan 48 EC) + one HW	0.48	Pre-plant

These herbicides provide very effective control of annual grasses and broad leaf weeds. The application of all pre-emergence herbicides must be done immediately after sowing when the field had enough moisture in order to get desirable effects. Mix these herbicides in 400- 500 lit of water per hectare and spray uniformly by using ûat fan or ûood zet nozzles, if weeds appear with the application of low dose of pendimethalin (0.45 kg/ha) give one hand hoeing. However, good crop stand integrated with chemical method, keeps the weeds under control throughout the growing season of crop.

Rice-Fallow Blackgram

i) Apply Imazethapyr @ 500ml ha^{-1} on 15-20DAS for Rice fallow Blackgram (Urdbean)

Oilseeds

Groundnut

i) Pre-sowing: Fluchloralin at 2.0 l/ha may be applied and incorporated.

ii) Pre-emergence: Fluchloralin 2.0 l/ha applied through flat fan nozzle with 900 lit of water/ha followed by irrigation. After 35-40 days one hand weeding may be given.

iii) Pre-emergence application of Metolachlor (2.0 l/ha) plus one hand weeding on 30 days after sowing is more profitable.

iv) Pre-plant incorporation of Trifluralin (Treflan 48 EC) 0.75 kg ai ha^{-1} + HW

v) Pre-emergence application of Pendimethalin @ 0.75 kg ai ha^{-1} (Stomp 30 EC)+ 1HW

vi) Pre-emergence application of Alachlor (Lasso 50 EC) @ 1.25 kg ai ha^{-1}+ HW

vii) In case no herbicide is applied two hand hoeing and weeding are given 20th and 40th day after sowing.

Gingelly

i) Weed and hoe on the 15th and 35th day of sowing.

ii) Apply Alachlor at 2.5 l/ha on 3 days after sowing and irrigate the crop immediately.

Sunflower

i) Apply Fluchloralin at 2.0 l/ha before sowing and incorporate or apply as pre-emergence spray on 3 days after sowing followed by irrigation or apply Pendimethalin (3.0 l/ha) as pre-emergence spray on 3 days after sowing. The spray of these herbicides has to be accomplished with Back-pack/Knapsack/Rocker sprayer fitted with flat fan nozzle using 900 lit of water/ha as spray fluid. All the herbicide application is to be followed by one late hand weeding 30-35 days after sowing.

ii) Hoe and hand weed on the 15th and 30th day of sowing and remove the weeds. Allow the weeds to dry for 2-3 days in the case of irrigated crop and then give irrigation.

Safflower

i) Hoe and hand weed on 25th and 40th day of sowing.

ii) Pendimethalin (Stomp 30 EC) @0.75 Kg a.i. ha^{-1}as pre-emergence spray

Castor

i) Hoe and weed on 20th and 40th day of sowing.

ii) Apply pre emergence herbicide Pendimethalin @ 3 lit/ha or Fluchloralin @ 2 lit/ha on 3 DAS followed by hand weeding twice on 20th & 40 DAS.

Linseeed

i) Isoproturon @ 0.94 kg a.i ha^{-1}as pre-emergence or Post-emergence (20-25 DAS)

Fibre Crops

Cotton

Cotton (*Gossypium hirsutum*) crop is severely infested with weeds due to wider row spacing of crop and favourable growing conditions for weeds prevailing during the crop growth season. All these weeds can be controlled either with mechanical means or with use of selective herbicides.

i) Trifluralin (Treflan 48 EC) 1.20 kg/ha as pre plant application or pendimethalin (Stomp 30 EC) 0.75 kg/ha or Fluchloralin 2.2 l/ha as pre emergence application using a hand operated sprayer fitted with deflecting or fan type nozzle keeps the crop free from weeds for 4-5 weeks. Sufficient moisture should be present in the soil at the time of herbicide application or irrigate immediately after application. Later on give hand weeding (or) directed spray of non-selective herbicides like paraquat (Gramoxone 24%) or glyphosate (Round up 41% / Glycel 41%) can be done 6-8 weeks after sowing cotton crop. Paraquat at 0.3 kg/ha controls only annual weeds whereas perennials regenerate within 5-7 days after application. However, application of glyphosate at 1.0 kg/ha takes care of all annual and perennial weeds. Use protective hood while spraying non-selective herbicides in order to avoid crop injury. Intercultural operation with tractor once or twice given at a time when crop is at knee high stage is also very useful.

Note: Do not use Diuron (Karmex) in sandy soil. Heavy rains after application of Karmex may adversely affect germination of cotton seeds.

ii) Hoe and hand weed between 18th to 20th day of sowing, if herbicide is not applied at the time of sowing followed by second hand weeding on 35 to 45 DAS.

Rice Fallow Cotton

i) Pre-emergence application of Fluchloralin 2.2 l/ha or Pendimethalin 3.3 l/ha ensures weed free condition for 40-45 days. This should be followed by one hand weeding and earthing up during 40-45 days. Fluchloralin need incorporation.

ii) Take up hoeing and weeding 20 days after sowing.

iii) Take up this operation when the top soil dries up comes to proper condition.

Rainfed Cotton

i Application of Fluchloralin 2.0 l/ha or Pendimethalin 3.3 l/ha or Thiobencarb 3.0 l/ha followed by hand weeding 40 days after crop emergence. At the time of herbicide application sufficient soil moisture must be there. Fluchloralin needs soil incorporation.

ii) If sufficient soil moisture is not available for applying herbicides hand weeding may be given at 15-20 days after crop emergence.

Sugar Crops

Sugarcane - Pure Crop

All weeds particularly perennials must be controlled by hand hoeing twice or thrice depending upon the level of weed infestations. Blind hoeing is quite helpful in controlling annual/perennial weeds and also improves germination of sugarcane crop. Later on one to two hoeing /interculture may be given when the crops is knee high stage.

i) Spray Atrazine 2 kg or Oxyflourfen 750 ml/ha (or) metribuzin (Sencor 70 WP) at 1.4 kg/ha or diuron (Klass 80 WP, Karmex 80 WP) at 1.6 kg/ha mixed in 900 lit of water as pre-emergence herbicide on 3rd day of planting, using deflector or fan type nozzle.

ii) If pre-emergence spray is not carried out, go for post-emergence spray of gramaxone 2.5lit + 2,4-d sodium salt 2.5 kg/ha in 900 lit of water on 21st day of planting or apply 10% Ammonium sulphate on 45th, 75th and 105th day after planting as directed spray.

iii) For the control of *Cyperus rotundus* (Dila/Motha) and broad leaf weeds 2,4-D, sodium salt at 1.6 kg a.i/ha can be sprayed as post-emergence. Few climbing weeds like *Ipomea sp*s. can also be controlled with 2,4-D Sodium salt at 1.6 kg/ha (or) 2, 4-D amine salt at 0.58 kg/ha as post-emergence.

iv) If the parasitic weed Striga is a problem, Post-emergence application of 2,4-D sodium salt 1.75 kg/ha in 650 lit of water/ha has to be sprayed. 2,4-D spraying should be avoided when neighbouring crop is cotton or bhendi or apply 20% urea for the control of Striga as directed spray.

v) If herbicide is not applied work the Junior-hoe along the ridges 25, 55, and 85 days after planting for removal of weeds and proper stirring. Remove the weeds along the furrows with hand hoe.

Sugarcane - Intercrop

i) Pre-emergence application of Thiobencarb 2.5 l/ha under intercropping system in sugarcane with soybean, black gram or ground nut gives effective weed control.

ii) Under inter-cropping situations, use herbicide which is selective to both crops. For example, if winter maize is inter-cropped in autumn cane, use atrazine at 0.75 kg/ha as pre-emergence application. Isoproturon at 0.94 kg/ha and 0.75 kg/ha can be used if wheat and raya are inter crops respectively.

Sugarcane-Ratoon Crop

i) Weeds from ratoon crop of sugarcane can also be controlled with the use of above herbicides provided these are applied after intercultural operations/ hand hoeing i.e. after removing the grown up weeds. However, atrazine is effective on emerged weeds provided these are at 2-3 leaf stage.

Integrated Weed Management in Different Cropping Systems

In **groundnut-maize** sequence, pre-emergence oxyûuorfen (0.125) or ûuchloralin (0.34 along with one manual weeding at 30 DAS for groundnut and blanket application of atrazine (0.25) followed by hoeing 30 DAS for maize.

In **transplanted rice-pulses and rice-rice (both transplanted)** sequence, pre-emergence application butachlor (1.25), thiobencarb (1.25), oxadiazon (0.5), pendimethalin (1.0), anilophos (0.4) alone or at 50 per cent of these herbicides plus 2, 4-DEE (0.5) tank mix followed by one weeding at 30-35 DAS for rice.

In rice-wheat rotation, application of butachlor or thiobencarb (1.5) as pre-emergence or each imposed with 2, 4-D (1.0) post-emergence at 20 DAS in drilled rice while isoproturon or oxadiazon (1.0) in wheat can be effective for controlling weeds without any residual effect on succeeding crops.

In **soybean-wheat rotation**, pre-emergence oxadiazon (0.6), butachlor (1.25), ûuchloralin (1.0) to soybean had no residual effect on succeeding crop of wheat.

Weed Control in Intercropping Systems

Herbicide identified for different crop combinations are:

Intercropping system	Promising herbicides
Wheat + gram	lsoproturon (1.0 kg)
Wheat + mustard	Isoproturon (1.0 kg)
Maize + Mung	Butachlor (1.25 kg)
Maize+Soybean	Alachlor (2.0 kg) or Pendimethalin (1.0 kg)
Maize + cowpea	Alachlor (1.25 kg)
Sorghum + Black gram	Pendimethalin (1.0 kg) or Alachlor (1.5 kg)
Sorghum + lablab	Pendimethalin (1.0 kg) or Dinitroamine (0.75 kg)
Sorghum + pigeonpea	Pendimethalin (1.0 kg)
	Ametryn (1.5 kg)
	Prometryn (1.0 kg)
	Terbutryn (1.0kg)
Pigeonpea + groundnut	Pendimethalin (1.0kg)
	Fluchloralin (1.0 kg)

Control of Parasitic Weeds

Parasitism is a biological mechanism to gain something for nothing. Parasitic weeds depend on host plant for mineral nutrients and destroy them subsequently. Important parasitic weeds in India are loranthus, dodder, orabanche and striga.

Loranthus (Bird Vine) (*Loranthus longiflorus*)

It is a semi stem parasite on mango, citrus, teak, casuarina and other trees. It occurs on a variety of hosts such as Albizzia, Acacia, and Melia. Occasionally, it causes considerable damage to pomegranate. Because of its dense presence on the trees, control is difficult. Hand picking the parasite in the early stage of its growth is the most common method of control. Spraying of 30 to 40 % diesel oil emulsion in soap water has also been found effective in eradicating the parasite.

Copper sulphate 7 g and 1.0 g fernoxone (2, 4-D) per tree is also found effective.

Cuscuta (Dodder) (*Cuscuta campestris, C. trifoli*)

It is total stem parasitic annual plant with twining thread like yellow or orange stems and leaves. The shoots of parasite twines the suitable host species closely, establish connection with it by producing numerous sucking organs called 'haustoria'. These organs extract water and nutrients from the host plant. If the parasite establishes a strong hold on the host, the latter is almost smothered and dies with the depletion of food. This parasitic weed affects legumes (Lucerne), niger and forest trees. Cuscuta does not parasitize any crop of grass family and hence they may be used in rotation. Control measures include:

- Planting crops other than legumes and ûax for several years to reduce weed seed in the soil,
- Application of CIPC at 6 to 8 kg/ha in lucerne crop infested with dodder,
- Application of paraquat at 1.0 kg/ha over infested Lucerne, and
- Use of pre-emergence (Chlorpropharn at 4.0 kg/ha) or post-emergence Pronamide application at 1.5 kg/ha 15 days after cuscuta germination.

Orabanche (*Broom Rape*) (*Orabanchecernua*)

It is a total parasite infesting tobacco and a number of solanaceous plants. It is brown or yellowish herb, without green leaves. This root parasitic weed entirely depends on the host plant (tobacco) for its growth and development. Its infestation results in stunted crop growth and very poor yield. It is endemic every year, but in certain years it attains epidemic proportion. Orobanche occurs at an intensity of 4.3 to 7.3 shoots per diseased plant. Growth inhibition and yield loss in tobacco is to the tune of 50% with an infestation of 7.3 shoots per host plant.

Hand removal is the quick method of control of orobanche but is very labourious and expensive. Trap crop can be very well used in rotation to control the weed. The trap crops are sunûower, sesame, cotton, soybean, and finger millet. These crops stimulate germination of the parasitic weed. But the infected weed dies either during the ûlamentous stage or during the subterranean button sized growth stage without any subsequent emergence into ûowering shoot. A long rotation with tobacco crop once in three years preceded by a chilli crop can reduce the loss due to the parasite. Application of kerosene at the tip of emerging orobanche arrests its growth. Though chemical weed control is not very effective, application of post-emergence herbicides 2,4-D sodium ethyl sulphate of alley alcohol at 0.1 to 0.2 % has given high percentage of control. Pre-emergence application of sodium 2, 4-D ethyl sulphate (sesone), post-emergence allyl alcohol (0.1 to 0.2 %) and directed post-emergence DNDC at 0.40 to 0.75 per cent can give good control. Lowland rice cultivation is also an effective method of controlling the orabanche.

Striga (Witch Weed)

The root parasite infests a number of crops Gramineae family particularly sorghum, maize and sugarcane. When the weed population is high, they completely suppress the crop growth, causing heavy yield loss.

This parasitic weed gives a serious problem in the whole of Asia and Africa. The yield loss ranges from 15 to 75% depending on the severity of infestation.

In India, ûve species of Striga are known to occur (Hosamani, 1978). They are: *S. lutia, S. angustifolia, S. densiûora, S. gesnerioides*, and *S. sulphurea*. Among them, *S. lutea*causes economic loss in agriculture particularly in crops like sorghum and millets. Striga control can be achieved by adopting different measures such as inducing the germination of the seeds, by rotating with trap crops, destroying the emerged weeds by tillage and with herbicides. Trap crops produce chemical stimulants to induce the germination of the weed seeds. Trap crops such as soybean, cowpea, groundnut, chickpea and pigeon pea can be used in rotation with sorghum, maize or millets. Post-emergence herbicides such as 2,4-D, MCPA at 1.0 kg/ha or paraquat can be used to kill the emerged weeds. Pre-emergence application of simazine or atrazine effectively controls the parasitic weed in maize, sorghum and sugarcane.

Control of Problematic Weeds

Nut sedge (*Cyperus rotundus* L.)

It is an obnoxious weed distributed throughout the tropical and temperate zones of the world.

Cultural method: Deep ploughing and exposure of the underground tubers and rhizomes to sun, removal of weeds through manual labourers is required. The tubers and rhizomes exposed to the solar heat may result in moisture depletion below the critical limit and they die untimely. Nutsedge is sensitive to light and could be suppressed in shade. A quick growing crop which will cover the canopy may the infestation satisfactorily. It has been noticed that by growing a green manure crop, daincha continuously for three years under daincha- potato- wheat sequence, the aerial shoots and tubers of the grass declined by 41.9 and 53.1% respectively.

Chemical method: Glyphosate, a non-selective herbicide @ 1 kg/ha either with 0.05% 2,4-D sodium salt or with 1% ammonium sulphate is more effective in the suppression of this weed. Application of MSMA/DSMA (4.5 kg ha^{-1}) twice at an interval of 10 days on standing nutsedge or 2, 4-D (2.0 kg ha^{-1}) repeated twice, glyphosate (0.8 to 1.5 kg ha^{-1}) on standing nutsedge/directed spray on plantation and orchard crops, with repeated spot sprayings will also suppress this weed.

Ipomea cornea: 2, 4-DEE (1.0 to 2.0 kg ha^{-1}) as post-emergence.

Water hyacinth (*Eichhornia crassipes*)

Manual removal of weed is the successful way of its control. Most commonly used herbicide is 2, 4-D at 1.2 to 8.0 kg/ha depending on weed mat density. To be effective, the spray liquid concentration of 2, 4-D should not be less than 0.3

per cent. Best time of application is summer, when the weed is growing actively. Combination of paraquat 1.0 kg/ha and amitrole 3.0 kg/ha or amitrole and fenac 0.1 kg/ha are very active.

Bermuda Grass (*Cynodon dactylon*)

Cultural methods such as summer ploughing, digging, ûooding, etc. keep the weed infestation under control. Most suited herbicides for the control of the weed are dalapon or TCA (2-4 kg/ha); Fluazi-fop-butyl or glyphosate (1.0 kg/ha).

Parthenium (*Parthenium hysterophorus*)

Cultural and physical control: In short statured crops and/or crops where lot of area has to be left around plants as their technical requirements for growing such plants e.g. orchards, plantation crops etc. Suitable cultural practices have to be adopted to remove the weeds before they start flowering. While adopting manual removal, care should be taken not to deploy persons who are susceptible to Parthenium allergy. Manual removal is generally adopted in gardens, flower beds, lawns and intensively cultivated agricultural fields. In fallow lands or common land belonging to panchayats, the weed removal should be adopted as a mass campaign involving voluntary organisations and school children. The mass removal method is best organised after the rain when the soil is moist enough to facilitate easy removal of Parthenium. In situations where ploughing is possible, the weed should be ploughed under before setting of flowers. Where ploughing is not possible, uprooting of this weed before flowering can also provide much relief. The workers engaged on this operation should avoid physical contact with the weed plant and should wear rubber gloves. Manual removal is an effective method but the persons engaged for uprooting should not be allergic to Parthenium. Alternatively, the plant should be cut with some sharp hand tool with a long handle before the ground level by avoiding contact with the plant. The cut plants should be stacked at one place and burnt. This operation should be adopted on a campaign basis throughout village/city/block/district/State. The government should direct the concerned agencies/organizations/owners of the land to follow strictly the control measures against the weed otherwise strict action would be taken against the irresponsible persons/organizations.

Exploiting plant competition: Using a harmless plant to displace a harmful plant is considered a novel approach. Few quick growing and commercially cultivated ûeld crops like bajra, sorghum and maize hardly allow Parthenium to grow. The potentiality of using a non-nitrogen flxing leguminous plant *Cassia uniflora* suppression of Parthenium through competitive suppression or displacement has been suggested (Singh, 1983). But this method of displacement of plant species sounds good for vast wasteland, and non-cropped border areas.*Cassia* sps. also has allelopathic action which hampers the growth of Parthenium. Several other plants like *Tephrosea purpurea*, *Amaranthus spinosus*, *Hyptis savolensis*, *Sida latifolia*, *Croton sparciûorus* and other species of *Cassia* have been found to suppress Parthenium (Anonymous, 1992).

Chemical control: A large number of chemicals such as paraquat, monosodium methane arsenate, disodium methane arsenate, atrazine, 2,4-D sodium salt, bromacil, glyphosate or sodium chloride kills the established. Fernoxone (80% sodium salt of 2,4-D) at 2.5 kg a.i./ha as a post emergent spray gave complete control of *P. hysterophorus* at all stages of its growth within a period of 20 days after the spray. For the control of Parthenium around Bangalore some individuals are using 20% saline water spray, or 5 cups of waste soap (the left over after washing clothes plus one cup of salt).

The herbicides/chemical that can also be used as post emergence application are atrazine (0.25 kg/ha + 2,4- D-1.25 kg/ha, paraquat 1 kg/ha, 2,4-D-1 kg/ha or 20% common salt spray.

After testing a number of herbicides, application of atrazine (Atrataf 50 WP) at the rate of 1.25.kg/ha or glyphosate 41 SL (Round up/Glycel) at 1.0 kg/ha have been recommended, which provides very effective control of Parthenium. Anyone of the herbicide must be dissolved in 500 litres of water and sprayed uniformly over the area of one hectare infested with this weed. Glyphosate is non-residual herbicide so any ornamental, vegetable, fruit or other economical plants can be sown when Parthenium plants dry up. However, residue of atrazine remains for a long time in the soil. For attaining good control, its stage of application is very important. These herbicides must be sprayed when weed plants are very young i.e. when leaves are tender and it has not flowered. On grown up plants, the commercial dose of Atrataf should be increased from 2.5 kg to 3.75 kg/ha. Application of atrazine should be done to moist soil i.e. after rainfall so that proper well kill can be obtained. The left over patches of Parthenium can be re sprayed after an interval of 15 to 21 days. Addition of ammonium sulphate (0.5%) or urea (0.5 to 1.0%) as tank mix application with glyphosate results in improved efficacy of this herbicide.

Biological control: Permission to import *Smicromyx lutulentus* Dietz. (Col., Curculionidae) and *Zygogramma bicolorata* Pallister (Col., Chrysomelidae) was given by the plant protection Advisor to the Government of India in March, 1983.

Control of Aquatic Weeds

Aquatic Weed Control

Aquatic weeds grow in water and complete life cycle, at least a part, in water: Water hyacinth, typha, water fern and hydrilla are the common aquatic weeds which need control measures. In recent years, rapid spread of aquatic weeds is creating serious socioeconomic problems.

Water Hyacinth

Water hyacinth (*Eichhornia crassipes*) is a perennial, free floating aquatic weed. The seeds germinate on the banks of water bodies and the seedling then move into the main water along with water waves. The mats are so strong as to even stop ships. Manual removal of weed is the successful way of its control. Most commonly used herbicide is 2, 4-D at 1.2 to 8.0 kg/ha depending on weed mat density. To be effective, the spray liquid concentration of 2, 4-D should not be less than 0.3 per cent. Best time of application is summer, when the weed is growing actively. Combination of paraquat 1.0 kg/ha and amitrole 3.0 kg/ha or amitrole and fenac 0.1 kg/ha are very active.

Typha spp. (Cattails)

Three species of typha (*T. latifolia, T. aungustifolia and T. elephantine*) are common in India. Manual weeding is difûcult because of its deep underground rhizomes. In shallow water areas, combination of dalapon 15 kg/ha and amitrole 3 kg/ha with 0.5 per cent surfactant can give good control. Diuron 15 kg/ha mixed with MSMA 2 kg/ha can give conspicuous control. A combined spray of 2, 4-D amine 1.5 kg/ha with aminotrizole 1.5 kg/ha is very effective.

Salvinia spp. (Water Fern)

Water fern (*Salvinia molesta*) is a free ûoating plant reproduced by offsets. Physical control by dredging with a rope is successful if care is taken not to miss plants. Granular C-288 30 kg/ha formulation (0.75 kg ai/ha) can completely control salvinia at low water level in rice ûeld or low lying areas without any injury to rice crop. Diquat, 2,4-D and mineral oils are effective against *Salvinia spp.* Diquat at 1.0 kg ha^{-1} was found destroy the plants in 5 days but dead mats were replaced soon by new ones. 2, 4-D was effective only at high rates of 25-40 kg ha^{-1}.

Hydrilla

Hydrilla verticillata is an annual submerged troublesome aquatic weed in calcareous water logged soils. Excellent control of hydrilla can be achieved with single application of tank mixed diquat and CSP (1.0 and 4 ppm) on the water surface. It is also highly susceptible to anhydrous ammonia and endothal. Common snail (*Marisa cornuarieties*) feed on hydrilla and destroys them.

Common aquatic algicides and herbicides

Name of the compound	Rate of application	Target plants	Toxicity
A. Algicides Copper sulphate penta hydrate	0.1 ppm 1.00 ppm 5.00 ppm	Water bloom common scum algae *Chara sp.*	MRL: 2.3 to 12.0 ppm in human drinking water. 100 ppm in animal drinking water
S. triazines Ametryn Terbutryn Simazine	0.05 to 0.1 ppm 0.5 to 1.25 ppm		Unsafe for treating irrigation and drinking water bodies
Dichlone	0.02 to 0.05 ppm 0.15 ppm	Water bloom Algal scums	Toxic to fish, treated water is unfit for drinking
Diuron	0.2 - 0.4 ppm	Macro algae	Accumulate in fish tissues
Hexa zinone	0.5 ppm	Macro algae	Toxic to fish

B. Aquatic herbicides

Acrolein (Acrylal dehyde)	15-80 ppm	Submersed weeds	Causes massive fish kill
Amitrole; Amitrole-T	5-15 kg ha^{-1} 1-1.5 kg ha^{-1}	*Typha* and *Phragmites sp.* *Eichhornia crassipes*	Safe to fish, water is unsafe for drinking
Ammonium Sulphamate (AMS)		Ditchbank brush killer	Safe to fish
Anhydrous ammonia	10 ppm	Submersed weeds (Hydrilla, Najas etc.) Water bloom	Destroy all vegetations as well as fish and other fauna. However, soon after the weed control, safe for varied uses.
Dichlobenil	5-10 kg ha^{-1} or 1 ppm	Emersed and Submersed macrophytes	Low toxicity to fish and wildlife.
Diquat and paraquat	0.25 to 1.5 ppm	Submersed weeds *Pistia, Lemna, Eichhornia, crassipes*	Unsafe for drinking, Diquat is safe to fish.
Diuron	10 kg ha^{-1} 1-2 ppm	Soil sterilant for treatment of water	Accumulate in fish tissues
Fenac	10-20 kg ha^{-1}	*Potamageton, Myriophyllum, Elodea, Najas* sp.	Unsuitable for drinking
Glyphosate	1-3 kg ha^{-1} in 300 l water	*Nymphaea, Phragmites communis*	Non residual
Silvex (fenoprop)	0.5 to 4.0 ppm	*Alteranthera philoxeroides, Nymphea, Nelumbo*	Unfit for irrigation and other uses.
Sodium arsenite	2-10 ppm	Submersed weeds and algae on confined water bodies	High mammalian toxicity.
2,4-D	1-8 kg ha^{-1} or 2 ppm	*Eichhornia crassipes, Myriophyllum, Najas* sp.	Unfit for irrigation, Toxicity to fish depend on the type of formulation

Localised control of aquatic weeds

Slow release dry materials of endothall (Hydout), slow release capsules of diquat (Envirocap D) and copper sulphate (Envirocap C) and slow release viscous formulations (Diquat alginate) are used for this purpose.

Methods of Application of Herbicides to Water Bodies

A. Ground application

The following devices can be used :

1. Boat sprayer,
2. Boat bailer,
3. Hoover craft,
4. Spray raft,
5. Weed wiper.

B. Aerial applicationUsing helicopter

Mostly done against *E. crassipes* and *Typha* spp.

Calculation of Herbicide Requirements

a) For the control of surface floating and emersed weeds, the water bodies are treated on the basis of area and hence the herbicide dose is given in terms of kg ha^{-1}.

b) For the control of submersed weeds in a confined water body it is necessary to calculate the quantity to produce a recommended concentration in specified column of water.

Q (in kg) = V (in cubic metre) x 0.001 x C (ppm w)

c) In flowing water bodies

On the basis of discharge in cubic meter per second = Wx D x U x 0.9

W = Average width (m)

D = Average depth (m)

U = Flow velocity of the ditch (m sec^{-1})

2.3

Herbicide History and Development

Herbicide History and Development

Common salt, ash, smelter waste, and like cheap materials have been used for centuries to control weeds on roadsides, fence rows, and pathways. But Chemical weed killing was born in 1896 when Bonnet, a French grape grower, observed that the Bordeaux mixture which he used against downy mildew of grape turned the leaves of *Sinapis arvensis* (a weed) black. Soon it was found that copper sulphate present in the Bordeaux mixture was responsible for its weed killing effect. Further tests in France, Germany, and U.S.A. established that copper sulphate could indeed be used for selective control of broadleaf weeds in small grains; wheat, barley, and oat. Between 1896 and 1908, several other inorganic salts, as also sulphuric acid, were found selective to small grain cereals. Of these, in Europe, a 6 to 10 per cent solution of sulphuric acid continued for considerable time as a satisfactory herbicide for broadleaf weed control in cereals, but it corroded the containers and sprayers. In the U.S.A., salts like sodium chlorate, carbon bisulphide, sodium arsenite, kainite, and calcium cyanamide were developed for non-selective eradication of perennial weeds, particularly, *Convolvulus arvensis* 'field bind weed' and *Cirsium arvense* 'Canada thistle'. In Hawaii, sodium arsenite solution was found highly effective for weed control in sugarcane plantations. Between 1930 and 1940, some boron compounds, thiocyanates, dinitro phenols (DNC and DNBP), ammonium sulphate, and certain mineral salts were developed in the west for selective and non-selective weed control. DNC and DNBP wereused extensively in Europe in small grains for broadleaf weed control, even though these chemicals dyed the hands and clothes of the users to yellow colour, almost permanently. Further milestones in weed control were introduction of organic chemical, 2-methyl-4, 6-dinitrophenoll (DMOC), in 1932 and discovery of hormone type selective weed killers. Selective residual chemicals such as substituted phenyl ureas and triazines, and non-residual chemicals diquat and paraquat are more recent discoveries.

The discovery of the herbicidal activity of 2, 4-D (2, 4-dichlorophenoxyacetic acid) first synthesized in 1941, triggered the development of modern herbicide technology. 2, 4-D proved to be an outstanding herbicide.

A real breakthrough in selective chemical weed control was achieved in 1945 with the announcement of simultaneous discovery of 2,4-D and MCPA in the U.S.A. and England,respectively, after the World War II. The discovery of these two sister compounds is said to have been affected much earlier than 1945 although it was then kept a war time secret till it was divulged later for agricultural use. While studying NAA for its growth regulatory effects on wheat in pot house, W. G. Templement of England observed in 1941 that it killed a few *Brassica kaber* 'wild mustard' plants growing as weeds in wheat pots. This induced him and W.A. Sexton at the Jealotts Hill Research Station, England to search for some more potent growth regulators than NAA for broadleaf weed control in small grains. IAA and NAA could not be used as field herbicides because these were found oxidised within the plants by oxidases enzymes. Their elaborate trials ûnally yielded 2,4-D and MCPA as potent growth regulator type of herbicides, the scientists were looking for. In the U.S.A., 2,4-D was discovered independently by PW. Zimmerman and A.E. Hitchkock at the Boyce Thompson Institute. Both 2,4-D and MCPA were found highly selective to cereals and phytotoxic to the broadleafs. They killed broadleaf weeds at very low rates. Besides, these compounds were (a) non-corrosive, (b) non-flammable, (c) non-explosive, and (d) non-toxic to man. Therefore, the two sister compounds rapidly replaced all the earlier used selective herbicides in small grain crops. 2,4-D stayed in the U.S.A. while MCPA continued in the European countries. In fact, the chemical control of weeds got a spurt with the introduction of 2, 4-D & MCPA.

The commercial success of 2, 4-D led to the development of other herbicides such as MCPA, silvex and 2, 4, 5-T, phenylurea herbicides such as monuron and linuron. It was during the 1950s and 1960s that the modern practices of using relatively low rates of synthetic herbicides for selective weed control in field crops was adopted in developed countries of the world.

Within ten years of discovery of 2, 4-D, the U.S.A. alone produced about 30,000 tonnes of this herbicide, annually. Since then 2, 4-D has been introduced in several other countries of the world, and today hundreds of thousand tonnes of 2, 4-D and MCPA are used round the globe to control herbaceous, broadleaf weeds in different situations.

In the agriculturally developed countries, herbicides form over 45% of the total pesticides used. Against this, in India, herbicides share only about 8% of the total pesticides consumed; the insecticides contributing to almost 77% of the total pesticide consumption. In absolute terms, in India we use an average of only about 40 g herbicide per ha per year, against 675-1350 g/ha/year used in many advanced countries. In Japan, as much as 5000g herbicide is used per ha, annually.

The spectacular acceptance of 2,4-D and MCPA for weed control in graminaceous crops in the world gave impetus to researchers to discover additional herbicides, resulting in a spate of new herbicides in certain agriculturally advanced countries. Today there are over 400 different herbicide molecules at global level, belonging to varied chemical families and groups; the latest being the sulfonylureas and their derivatives. Over 100 private companies are involved in research and development of these herbicide compounds, largely in the western world. The research is ongoing to invent new herbicide structures to deal with the ever emerging new weed problems.

The introduction of glyphosate a non-selective herbicide in the late 1970's provided outstanding control of most perennial grasses and many perennial broadleaf weeds. In modern herbicide development, the discovery of novel organic compounds that exhibit phytotoxic properties often lead to the synthesis of related compounds, where chemist attempts to optimize herbicidal activity of the original compound by altering or modifying its chemical structure. Herbicide development in the 1980s was marked by the introduction of selective post-emergence treatments in major crops such as sulfonyl ureas, imidazolinones and aryloxy phenoxy propionate not only provide excellent selectivity but are used at extremely low dosage. Consequently there are often several herbicides developed within a chemical family that are structurally related and have essentially the same mode of action in plants. Let us see the structures of important herbicides.

Presently in India there are 42 herbicides that are registered under the Insecticide Act of 1968 of Govt. of India. Besides there are some 10 new herbicides under various levels of testing and use. Wheat and rice are the major consumers of herbicides in the country with 60% and 20% share, respectively. Soybean is another important crop with a herbicide consumption share of about 4%. The rest of the about ten thousand ton of herbicide produced in India each year is consumed by other crops, particularly tea. In all, approximately 20 m ha of cultivated land, which is about 10% of our gross cultivated area, is treated annually with herbicides.

Procedure Outline and Costs of Inventing New Herbicides

The invention of new herbicides with novel modes of actions is a priority assignment in plant protection research. Almost all herbicides in the world are invented by private companies in the ûrst instance, but often the Universities and public and Government research groups are latter involved in their further testing and development. The invention of each successful herbicide requires, on an average, testing of over 7000 compounds, though in most cases a company starts with 10,000 molecules. A common means to getting to such a huge number of new chemical structures is to synthesize homologues and analogues of a proven phyto-active material. But, once in a while completely new chemical structures are also tried. Often plant's native chemicals, particularly, the hormone and enzyme structures, form the basis of developing new herbicides. From the planned synthesis of new structures, and their numerous variants, come candidate herbicides. The large companies, as a routine, screen these chemicals in greenhouses for their biological and pesticidal activity which suddenly reduces their number to about 40 or so. The remainder chemical structures are ûled for future reference.

For active compounds identified in greenhouse screenings, the crucial point is to tread the most efficient path in determining herbicides' target sites, chances of success, and time and research costs.

The selected compounds are then subjected to more extensive laboratory and small field plot tests, using a large number of plant species. Of these compounds, one or two are ûnally selected for their introduction in field studies on manufacturers' own farms located at different places in the country. The chemical process research and pilot plant production are then taken up. Concurrently, analytical techniques are developed to detect any possible residues of candidate compounds in the treated crops. Also, feeding tests on animals are conducted. The bio-chemical studies on the fate of compounds in plants and soils are also made. At this stage the herbicide under investigation is offered to the Government laboratories and experiment stations as test compound for their independent evaluation.

If the data from the above evaluation schemes are still favourable, an application is made to the appropriate Government agencies for temporary permit to allow ûeld tests in a variety of geographic conditions. Finally, when a compound has passed through all these tests successfully, a registration from the Government is sought. For instance, in the U.S.A., Pesticide Regulation Division (PRD) and the Environmental Pollution Agency (EPA) of USDA are involved in pesticide clearances.

The above complex procedure of discovering herbicides calls for the combined efforts of the chemists, physiologists, agronomists, toxicologists, and environmental biologists for an average period of six and a half years. And during this time the company has spent a few million dollars. These figures, incidentally are tending to increase from year to year.

Objectives and Scope of Herbicide Usage

The primary objective of introducing herbicides in any country has been to replace the torturous and back-breaking, manual weeding and to give the farmer the much needed leisure, of which he has been deprived since ages. The spared labour could be utilized more effectively for bringing additional land under cultivation and improving the irrigation, fertilization, and other pesticide application practices, enhancing thus the overall agricultural production in the country. The swelling costs of hired labour and limited supply of fuel for mechanical weed control are two other reasons for introducing herbicides in agriculture. The herbicides are not aimed at substituting the physical, biological, or good crop husbandry methods of weed control. They are meant to bridge the gaps in these methods and act as added production tools in agriculture. Practical weed control is a combined effect of good crop husbandry, herbicides, and other techniques. Herbicides have ample scope in vegetation management, both on crop lands and non-crop lands. On crop lands, herbicides are used as aids in seedbed preparation for destroying the established weeds before they have a chance to infest the crops. Later, these can be used to raise weed free crops with several advantages. Outside the crop field, herbicides can be employed to maintain weed sanitation as part of weed prevention work for the main ûelds. Also, herbicides form useful tools in limiting the unwanted vegetation in industrial areas, roadsides, along the rails, and in several others like situations. In water bodies too, herbicides can be efficiently employed to free them of their weedy vegetation that hinder proper water use. In summary, where there is need to manage vegetation efficiently, there is scope of using herbicides, along with the other inputs.

2.4

Classification and Characteristics of Herbicides

History on Classification of Herbicides

Copper sulphate ($CuSO_4$) was the earliest inorganic salt used as selective herbicide for controlling dicot weeds in cereal crops during 1896 in France, Germany, and USA. Sulphuric acid (H_2SO_4) was used as herbicide in cereals during 1909. In selective killing action depends on the fact that it runs off the leaves of the cereals and scorches rough surfaces of certain seeds. Sodium chlorate ($NaClO_3$) under the name mort-herbs is used in France for killing deep rooted perennials. Another non-selective herbicide, ammonium sulphamate ($NH_4SO_3NH_2$) was introduced in 1945 as **ammate** for the control of woody plants. Organic herbicides are more effective and selective.

Classification of Herbicides

Classification of herbicides is modified periodically as and when new herbicides are developed. Herbicides are classiûed based on their method of use (soil and foliage applied), mode of action (systemic or translocated and non-systemic or contact herbicide) and chemical key structure (aliphatic-aromatic, dinitrophenal, etc. Some herbicide molecules contain more than one classifying key structure and they are put into the group most likely to be contributing activity.

Any approved herbicide is known by three names: common name, chemical name and trade name. Common name of herbicide is a technically accepted abbreviation of its chemical name. Chemical name reads out its full molecular structure. Trade names refer to their formulations offered by the chemical companies for the consumers.

Comprehensive herbicide classiûcation based on chemical structure is given below:

1. On the Basis of Chemical Structure

1. ***Inorganic Herbicides*:** First chemical used, Arsenic, Sod Sulphuric Acid Sodium arsenate, Sod Chlorate, Borax, Copper Sulphate, Copper nitrate etc.
2. ***Organic Herbicides :*** 16 or 17 groups:

Herbicides and its Trade Name

Group	Herbicides
1. Aliphatic	Dalapon, TCA, Aerolein, CH_3 Br
2. Amides and Acetamides	All-chlor (like Butachlor, Alchlor (Lasso), Propachlor, Propanil 2,3,6-TBA, Dicamba, Tricamba, Chloramben
3. Baenjoics	Paraquat, Diquat (contact)
4. Bipyridiliums	Propham, barban, dichlormate
5. Carbamates	Butylate, Thiobencarb or benthiucarb
6. Thiocarbamates	CDEC, Metham
7. Dithiocarbamates	Bromoxynil, Dichlobenil
8. Nitriles	Fluchloralin (Basalin), Pendimethalin, Trifluralin, Nitralin
9. Dinitroanilines	Diniseb, DNOC, PCP
10. Phenols	2,4-D, 2,4,5-T, MCPA, MCPB, 2,4-DB, Dichlorprop
11. Phenoxys	Atrazine, Propazine, Simazine (soil applied)
12. Triazines	All-ron (like Diuron, Monuron, lsoproturon)
13. Ureas	Bromacil, Terbacil
14. Uracils	Nitrofen (Toke-25)
15. Diphenyl ethers	Picloram, pyrazon
16. Others	

2. On the basis of Selectivity

1. ***Selective herbicides:*** Such herbicides kills only target weeds e.g. Simazine, Atrazine, 2, 4-D,MCPa, Butachlor, Pendimethalin, Fluchloralin, Isoproturon etc.
2. ***Non-selective herbicide:*** Kills all vegetation (crop + weed both) when comes in contact e.g. Diquat and Paraquat. However, selectivity depends on dosage of herbicide.

3. On Translocation

1. ***Systemic Herbicide:*** Such herbicides move within the weed either through xylem or phloem and thus affect the whole system like photosynthesis and respiration. Most of the systemic herbicides are selective at recommended dose e. g. propanil, 2, 4-D, Atrazine, Simazine.
2. ***Contact herbicide:*** Kills the vegetation either weeds or crops when it comes in contact. It kills the part of the plant that is in contact with herbicide e. g. Paraquat and Diquat

4. On the Basis of Time of Application

1. ***Pre-plant applied:*** Such herbicides are applied before planting of crop in field e.g.Fluchloralin and Alachlor.
2. ***Pre-emergence:*** Applied before emergence of weeds but post-emergence of crops e.g. only selective herbicides.
3. ***Post-emergence:*** Applied after the emergence of weed (and also crop) e.g. 2, 4-D, Propanil,Diquat, Paraquat, Dalapon.

5. On the Basis of Method of Application

1. ***Foliage:*** It may be either contact or translocated.
2. ***Soil applied:*** Either selective or nonselective.
3. ***Aquatic application:*** E.g. Copper sulphate, 2, 4-D.

Classification of herbicide (organic compounds) based on chemical nature is presented in Table. 2.4.1

1. Synergistic Effect

a) Atrazine + 2,4-D (Low concentration)

b) Paraquat (low) + Pentachlor

c) Atrazine + Alachor (widely used in cotton)

2. **Antagonistic Effect :** Generally contact + systemic herbicide combination show antagonistic effect

a) Dalapon + Atrazine

b) TCA + 2,4-D

c) TCA + MCPA

- **Active ingredient (a.i.):** Chemical in commercial product that is directly responsible for the herbicidal activity is called active ingredient. Generally, expressed as per cent by weight or by volume.

  $$\text{The amount of commercial product} = \frac{\text{Recommended dose}}{\text{Product to be required}} \times 100 a.i.$$

- **Acid Equivalent (a.e.) :** Some herbicides are active organic acids like phenoxy acetic acid, piclorarn and chloramen. But many of these are generally supplied in the form of their salts and esters e.g. instead of 2,4-D in acid form, sodium salt or amine salt or ester form is used. The theoretical yield of parent acid from such a herbicide formulation is called its acid equivalent. The acid equivalent of sodium salt of 2, 4-D is 92.5%. It means 2, 4-D id 92.5% in sodium salt of 2, 4-D. The a.e. is less than a.i.

Table 2.4.1: Classification of herbicide based on chemical nature

Group	Common herbicide	Time of application	Rate	Crops/weed species
Acetamides	Acetochlor	PPI or PPE	1-5-3.0 kg/ha	Soybean, Maize
	Alachlor	OOI, PRE or E.Post	2.0-4.0 kg/ha	Maize, sorghum, groundnut, soybean, cotton
	Butachlor	PRE	1.0-3.0 kg/ha	Rice, groundnut, soybean
	Metalochlor	E-PPS, PPI, PRE or E-Post, POST	1.5-4.0 kg/ha	Maize, cotton, groundnut
	Propanil	POST	3.0-5.0 kg/ha	Rice, wheat
Aliphatic	Aerolin	-	15 g/100 lit	Floating weeds
Arsenics	DSMA	POST	2.5-5.0 kg/ha	Cotton, non-crop areas
	MSMA	POST	2.5 kg/ha	Cotton, non-crop areas
Benzamides	Isoxaben	PRE, POST or PPI	1.5-2.0 kg/ha	Tobacco
Benoics	Dicamba	PPS, PRE or POST	250.50	Maize. sorghum, wheat
Bipyridiliums	Diquat	POST	2.0-4.0 kg/ha	Aquatic weeds and desiccant in cotton
	Paraquat	POST	1.0-3.0 kg/ha	Non crop areas and plantation crops
Carbamates	Asulam	POST	1.0-3.0 kg/ha	Sugarcane, non crop areas
	Desmedipham	POST	0.8-1.5 kg/ha	Sugarbeet
Cineoles	Cinemethylin	PPI or PRE	0.5-1.5 kg/ha	Soybean, cotton, groundnut
Cyclohexanediones (Cyclohexenones)	Sethoxydim	POST	0.10-0.5 kg/ha	Cotton, sunflower
Dinitro anilines	Fluchloralin	PRE	1.0-4.0 kg/ha	Groundnutr. pulses
Pendimethalin		PRE or E-POST	0.5-2.0 kg/ha	Maize, sorghum, wheat, sunflower, cotton, sugarcane
	Trifluralin	PPI	0.5-1.0 kg/ha	Maize, soybean, sunflower, castor,safflower
Diphenylethers	Oxyflurfen	POST	0.25-2.0 kg/ha	Soybean, groundnut
Imida zolidinones	Buthidazole	PRE or POST		Maize, wheat
Imida zolinones	Imazethaphyr	POST	50-70 g/ha	Soybean
	Imazamethabenz	POST	20-40 g/ha	Sunflower
Inines	CGA-248757	POST	3-15 g/ha	Maize, soybean
Isoxazolidinones	Clomazone	PPI or PRE	0.56-1.7 kg/ha	Soybean
Nitriles	Bromoxynil Dichlobenil	POST POST	0.2-0.56 kg/ha 4.0-8.0 kg/ha	Wheat, barley, oats, maize, sorghum Forest nurseries
	Loxynil	POST	0.5-1.0 kg/ha	Wheat, barley
Oxa diazoles	Oxadiazon	PRE	0.5-4.0 kg/ha	Rice, soybean, groundnut, cotton, sugarcane

Oxa diazolidines	Methazole	PPI, PRE or POST	0.5-1.5 kg/ha	Sorghum, maize, wheat, soybean, cotton
Phenols	Dinoseb	POST	0.75-1.2 kg/ha	Sorghum, maize, wheat, soybean, cotton
Phenoxy acetics	2,4 -D	POST	0.26-2.24 kg/ha	Cereal crops, sugarcane, non crop areas
	MCPA	POST	0.25-1.5 kg/ha	Cereal crops, legume
Phenoxy butyrics	2,4-DB	POST	0.20-0.45 kg/ha	Forage, legume crop
Aryloxy phenoxy Proplonics	Fluazitos-P	POST	50-200 g/ha	Wheat, soybean, cotton
N-henylphthalimides	Flumoclorac	POST	30-80 g/ha	Maize, soybean
Phenyl pyridazines	Sulfentrazones	PPI or PRE	60-90 g/ha	Soybean
Phthalamates	Naptalam	PRE or POST	2.0-4.0 kg/ha	Groundnut, soybean
Pyrazoliums	Difenoqual	POST	0.7-1.0 kg/ha	Wild oats
Puridazinones	Picloram	POST	0.15-1.0 kg/ha	Non-crop areas
Pyridines	Dithiopyr	PRE	0.25-0.5 kg/ha	Rice
Pyridinones	Fluridone	-	0.06-0.09 mg/l	Aquatic weeds
Pyrimidinyl Thiobenzoates	Pyrithiobac	POST	0.07 kg/ha	Cotton
Quinolinecarxo	Quincloracxylic acids	PRE, late PRE or Early-POST	0.25-0.5 kg/ha	Rice
Sulfonylureas	Chlorsulfuron Sulfometuron	PRE or POST PRE or POST	10-20 g/ha 50-200 g/ha	Wheat Non crop areas
Thiocarbamates	Butylate Diallate EPTC Thiobencarb	PPI PPI PPI PRE	3.0-6.0 kg/ha 2.0-5.0 kg/ha 2.0-5.0kg/ha 1.0-1.5 kg/ha	Maize Wheat, maize, soybean, oilseed Maize, sunflower, cotton Irrigated rice
Triazines	Ametryn Atrazine Prometryn Simazine	POST PRE PRE PRE	0.5-2.0 1.0-2.0 0.5-3.0 1.0-4.0	Sugarcane Maize, soybean, sugarcane Cotton, soybean, non crop areas Maize, sugarcane
Triazinones	Metribuzin	PPI, PRE or POST	0.25-0.80	Maize, soybean, sugarcane
Triazoles Trizolo pyrimidine	Amitrole	POST	2.0-10.0	Non crop areas
Sulfonanilidea	Flumetsulam	PRE, PPI or PPS	50-75 g/ha	Maize, soybean
Uracils	Bromacil	PRE	1.5-5.0	Non crop areas
Ureas	Diuron Flumeturon	PPI PRE or POST	0.2-0.4 1.0-1.5	Maize, sorghum Sorghum, cotton, sugarcane
Unclassified	Bensulide Glyphosate	PPI or PRE POST	5.6-6.7 0.5-4.0	Vegetable crops Perennial weeds

(PPI= pre planting, PPS= pre planting surface, E-pps+ early pre planting surface, PPI= pre planting incorporation, PRE= pre emergence, POST=post emergence, E-POST= early post emergence)

Source: Rao, 2000

S.No.	Short Name	Full Name
1.	2,4-D	2, 4-dichlorophenoxy acetic acid
2.	2,4-DB	2, 4-dichlorophenoxy butyric acid
3.	2,4, %-T	2, 4, 5-trichlorophenoxy acetic acid
4.	TCA	Trichloro acetic acid
5.	Propanil (Stam F-34, Rogue)	3,4- dichloro proplonanilide
6.	MCPA	2-methyl, 4-chlorophenoxy acetic acid
7.	MCPB	2-methyl, 4-chlorophenoxy butyric acid
8.	Dalapon	2,2-dichloroproplonic acid (sod . salt)
9.	IPC (propham)	Isopropyl N-phenyl carbamate

Organic Herbicides

Organic herbicides are more effective and selective. Comprehensive herbicide classification based on chemical structure is given below in Table 2.4.2.

Table: 2.4.2 Comprehensive herbicide classification on chemical structure

S.No.	Chemical group/ common name	Trade name	Structural formul a
I.	**Aliphabatic (Halo-aliphatic acids)**		
	Dalapon TCA	Dowpon Varitox	2,2-dichloroproplonic acid Trichloro acetic acid
II.	**a) Anilides and amides**		
	Alachlor	Lasso	2-chloro-2,6-diethyl-N (methoxymethyl) acetanilide
	Butachlor	Machete	N-(butoxy methyl)-2-chloro, 6-ehtyl acetanilide
	CDAA (allidochlor)	Randox	N,N-diallyl-2-chloroacetamide
	Propachlor	Ramrod	2-chloro-N-Isopropyl acetanilide
	Propanil	Slam F-24, Rogue	3,4-dichloro propioionanilide
	Pronamide	Kerb	acetanilide 3,5-dichloro (N-1, 1-dimethyl-2-propynyl)
	Naptalam	Alanap	N-1-naphthylamic acid
ll. (b)	**Acryldehyde**		
	Acrolein	Aqualin	= 2-propanol or acryladehyde
III.	**Anilines and Nitro-phenols**		
i)	Denitramine	Cobex	2,6-dinilro-3, amino-4-tri-fluoromethyl-NN-diethyl-aniline
	Nltralin	Planavin	4-(methyl-sulphonyl)-2,6-dinitro NN- dipropy pynoline
	Penexalin	Stamp	N-(ethylpropy)-3,4-dlmethyl-2,6-Dinitro benzamine
	Tri?uralin	Treflan	2,6-dinitro-NN-depropyl-4-tri?uron Methyll aniline
	Fluchloralin	Basalin	N-propyl-N-(2, chloroethyl-2,6-Dilnitro tri?uoroin-ethyl-aniline
ii)	DNOC	Sinox	2-methyl-4,6-dinitrophenol

	Dlnoseb	Dow-weed killer Pre emergence	4,6-dinltro-2,5-buiylphenol
	PCP	Durotox	Pentachlorophenol
	Nitrofen	TOK E-25	2,4-dichlorophenyl-nilro-phenyl ether
	Fluorodifen	Soyex, Preforen	2,4-dinitro-4-trifluoro-methyl-diphenyl ether
IV. Arsenicals (Organo -arsenic Compounds)			
	DSMA	Ansar184, Chipco crab kleen	Disodium methyl arsenate
	MAA	Ansar, Weed - ERad	Methane arsenic acid
	MSMA	Ansar-529, Daconate	Mono-sodium methyl arsenate
V. Benzoics and Phynyl acetic acid			
	Chloramben	Amiben, Vegiben	3, amino-2,5-dichlorobenzoic acid
	Dicamba	Banvel-D	2-methoxy-3, 6-dichlorobenzoic acid
	Fenac	Fenac	2,3,6-lrichlorophenylacetic acid
	Oxyflourfen	Goal	2-chloro-1 (3 ethoxy-4-nitrophenoxy- (trifluor-mrthyl) benzene
VI. Carbamates and Thio-carbamate			
(a)	Carbamate		
i)	Alkyl Dichlomate	Sirmate	3,4-dichlombenzyl methyl carbamate
ii)	Aryl Asulam	Auslox	Methyl sulfanilyl carbamate
	Barban	Carbyne	4-chloro-2-butynyl m-chloro-Carbanilate
	Chlorpropham	ClPC	Isopropyl m-chloro carbanilate
	Propham	lPC	Isopropyl carbamate
(b)	Thio-carbamates		
	Benthiocarb	Saturn, Bolero	S-(4-chlorobenzyl) N,N-diethyl Thio- carbamate
	EPTC	EPTAM	S-ethyl dipropyl thio carbamate
	Diallate	Avadex	S-(2,3-dichloroallyl) di-iso-propyl-thio carbamate
	Triallate	Avadex BW	S-(2,3,3-trichloro allyl di-iso-propyl thio carbamate
	CDEC	Vegadex	2-chloro allyl diethyl dithio - Carbamate
	Molinate	Ordram-71	S-ethyl hexahydro H-azapine 1carbothionate
	Bentazon	BesagranBSA-3510	3-isopropyl-2,1,3-benzo-thiadiazon-4,1,2, 2-dioxide

	Carbo thionate SWEP Glycine derivatives Methyl 3,4- dichloro carbonilate Glyphosate	Non -0573	N (Phosphono methyl) Glycine
VII. Heterocyclic Compounds			
(a)	Bipyridylium		
	Diquat	Round-up, Reglone	1,1-ethyllene-2, 2 bipyridylum dibromide
	Paraquat	Gramoxone	1,1-dimethyl -4, 4-bipyridylium dibromide
(b)	Pyridines	Tordon	4-amino-3,5,6-trichloro piclonic acid
	Pichloram		
(c)	Pyridazines Maleic hydrazide (MH)	Retard	1,2-dihydropyridazine-3,6-dione
(d)	Pyrimidines		
	Bromacil	Hyvar-x, Urox	5-bromo-6-methyl1-3, 5-butyluracil
	Terbasil	Sinbar	5-chloro-6-methyl 3-T butyluracil
(e)	Triazines		
	Atrazine	Atrataf, Residox Gesaprin	2-chloro-4(ethyl amino)-6-lsopropyl amino)-S-uiazine
	Simazine	Gesatop, prinlep	2, chloro-4, 6-bis (ethyl amino)-S-triazine
	Propazine	Gesamil, Milogard	-chloro-4,6-bis (isopropyl Amino)-S- triazine
	Ametryne	Gesapax	2-(ethylamino)-4-(isopropyl-amino) -6- (methyl lithio)-S-triazine
	Prometone	Pramitol	2,4-bis (isopropy|amino)-6-methoxy-S- triazine
	Prometryne	Gesagard, Caprol	2,4-bis (isopropyl amino)-6-Methoxy-S- triazine
	Terbutryne	lgran	(tert-butyiamino)-4 (ethyl amino)-6- (methyltho)~S-triazine
	Metribuzin	Sencor, lexone	4-amino-6-tert-butyl 3-3-(methyl-Thio)-S- triazine
	Metribuzin	Sencor, Lexone Bey-94337	4-amino-6-tert-butyl 3-(methyl thio)-S- triazine-5(4H) one
(f)	Misc. heterocyclic Compounds		
	Aminotriazone		
	Amitrole	Weedazol, Amizol	3, amino-1,2,4-triazole
	Endothall	Endothall	3, 6-endoxohexa hydrophthalic Acid

	Oxadiazon	Ronstar	2-butyl-4-(2,4-dichloro-5-isopropyl Oxyphenol)-1,3,4-oxadiazolin 5 one
	Pyrazon	Pyramin	5-amino-4-chloro-2-phenyl 3 (2H) pyridazinone
VIII. Hormone type Herbicides			
(a)	Phenoxy acetic acids		
	2,4-D	Ethyl ester, Weedone Sodium Femoxone, Weedox, Tafacide, Amino - Dedweed-40 Special formulation-Decamine	(2,4-dichloro phenoxy) acetic acid
	2,4,5-T	Brush killer	(2,4,5 reichloro phenoxy) acetic acid
	MCPA	Chiptox	(4-chloro-2-methyl phenoxy) acetic acid
(b)	Phenoxy propionic acid		
	Dichlorprop	2,4-DP	(2,4-dichlomphenoxy) propionic acid
	Silvex	Weedone, 2,4,5 - TB Kuron Pheniorop	2(2,4,5-trichlomphenoxy) propionic acid
	Mecoprop	MCPP, Lomox - Plus	(4-chloro-2-methyl phenoxy) propionic acid
(c)	Y-Phenoxy butyric acid		
	2,4-DB	Butaxone	(2,4-dichloro phenoxy) butyric acid
	MCPB	Tropotox, Control, Thistrol	(4-chloro-2-methyl -phenoxy) Butyric acid
(d)	Misc.		
	2,4-D EP	Falone	Tris{2-(2-4-dichloro penoxy)} ethyl Phosphate
	@-$-DES	Sesone	Sodium, 2-(2-4-dichlorophenoxy) Ethyl- sulphate
	IX. Nitriles		
	Bromoxynil	Buctril	4-hydroxy-3,5-dibmmobenzonitrile
	Dichlobenil	Casoron	2,6_dichloro benzonitrile
	loxynil	Actril	4-hydroxy-3,5-di-iodobenzonitrile

X. Substituted Ureas			
(a)	Alkyl		
	Chloroxuron	Tenoran	3-(p-(p-chloro phenoxy) phenyl)1,1- Dimethyl urea
	Diuron	Karmex	1 3-(3,4-dichlorophenyl) dimethyl urea
	Fenuron TCA	Urab	1,1-dimethyl -3-phenyl-ureamon (trichloro acetale)
	Fluometuron	Cotoran, Lanex	1,1-dimelhyl-3-(a,a,a-tri?uoro m-Totyl) urea
	Monuron	Telvar	3(p-chlomphenyl)-1, 1-dimethyl Urea
	Neburon	Bonus	3,4-dichlorophenylbutyl, mrthyl Urea
(b)	Alkoxy		
	Linuron	Lorox, Alfalon	3-(3,4-dichlorophenyl)-1, methoxy-1 Methyl urea
	Chlorbromuron	Bromex, Maioron	3-(4-bromo-3-chlorophenyl)-1-Methylurea
XI. Unclassified			
	C-288-Methazole	Probe	2(3,4-dichlorophenyl)-4-methyl-1, Oxadiazolidine-3,5-dione
	Per?uidone	Destun	4-(phenyl sulphonyl)-1, 1, 1-Tri?uoro methyl sulphono-0-toluidide

Mechanism of Action of Herbicides

1. Alliphatics

a) Dalapon

i) Causes leaf chlorosis and necrosis.

ii) Arrests mitotic activity at prophase.

iii) Affects metabolic processes of plant

2. Amides

a) Alachlor, Propachlor and Prynachlor.

i) When applied pre-emergence inhibit seed germination.

ii) Prevent formation of m-RNA-ribosome complex

iii) Activation of amino acid to form amino-acyl-tRNA and thus inhibit protein synthesis.

b) Propanil

i) Causes chlorosis and necrosis.

ii) Inhibits radical elongation.

iii) Destroys the permeability of root membrane.

3. Arsenicals

i) First symptom is chlorosis.

ii) Cessation of growth and gradual browning.

iii Cause root plasmolysis and leaf wilting leading to necrosis of leaf tips and margins.

4. Benzoic and phenyl acetic acids

a) Chloramben

i) Inhibit elongation and growth of roots and shoots.

ii) Inhibit oxygen uptake.

b) Dicarnba

i) Causes growth proliferation

c) Trichlorobenzoic acid (TBA)

i) Inhibits apical meristems of dicotyledons.

5. Bipyridliums

a) Diquat and paraquat

i) Cause wilting rapid dissication of the foliage.

ii) Affect the plant surface through their ability to form toxic free radicals.

6. Carbamates

a) Asulam

i) Interfere with RNA and protein synthesis

b) Barban

i) Causes swelling of the apical growing points.

ii) Distorition of the meristem.

iii) Failure of auxiliary buds to develop.

c) Chloroprpham and Prophan

i) Inhibit mitosis and thus cell division.

7. Dinitroanilines

i) Inhibit both root and shoot growth and development.

ii) Primary mechanism of action is nucleic acid inhibition.

8. Nitriles

a) Bromoxynil and ioxynil

i) Necrosis and extensive destruction of leaf tissue.

ii) Uncoupling oxidative and photosynthetic phosphorylation.

9. Phenols

i) Green tissues turn brown and desiccate.

ii) Inhibition of oxidative phosphorylation.

10. Phenoxy acids (auxin type herbicides)

a) Phytotoxic symptoms are:

i) Epinasty swelling, twisting and bending of the treated area.

ii) Plants stop growing.

iii) Meristematic cells stop dividing.

iv) Young leaves stop expanding.

11. Thiocarbamates

i) Inhibit shoot and root growth of grassy weeds.

ii) Inhibit early seedling growth.

iii) Diallate and triallate caused mitotic abnormalities.

12. Triazines

i) Atrazines cause degradation of chloroplasts and also cause swelling.

ii) Triazines cause inhibition of photosynthesis and nitrite reduction.

iii) All triazine herbicides inhibited state 3 respiration most and cyclic photophosphorylation.

iv) Triazines also have growth stimulant effect on plants.

13. Triazoles

i) Development of albino leaves and shoots.

ii) Amitrole cause destruction of chlorophyll and impairment Chlorophyll development.

14. Uracils

i) Inhibit root growth.

ii) Inhibit development of the chloroplast grana and fret system in leaves.

15. Ureas

i) Cause chlorosis, growth retardation and ultimately death.

ii) Monouron produces abnormal meiotic cells in root tips.

iii) Substituted ureas inhibit the Hill reaction in photosynthesis.

Formulae of Important Herbicides

The commonly used herbicides are organic in nature. The herbicides depending upon their chemical composition are subdivided into different groups as follows:

Aliphatics

- Dalapon: 2,2-dichloropropionic acid

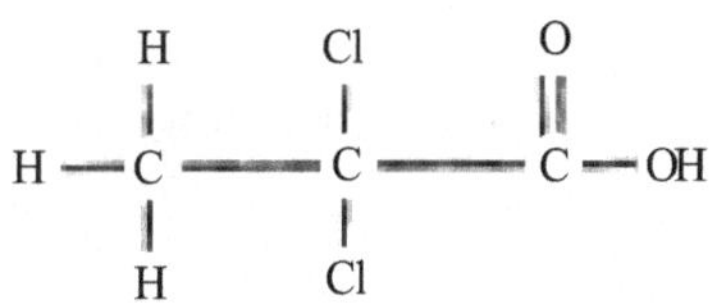

- TCA: Trichloro acetic acid

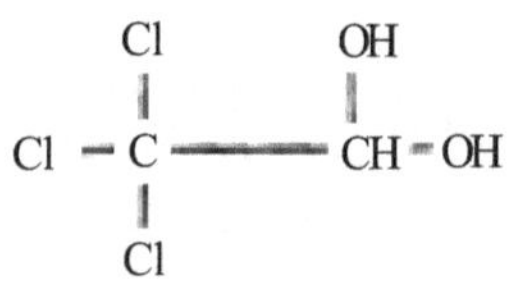

Amides

- Alochlor: 2-chloro-2',6'-diethyl -N-(methoxymethyl) acetanilide

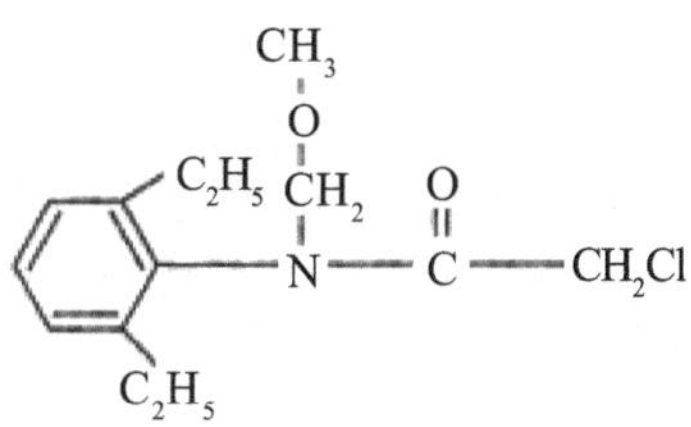

- Butachlor: N-(butoxy methyl)-2-chloro-2'6'-diethyl acetanilide

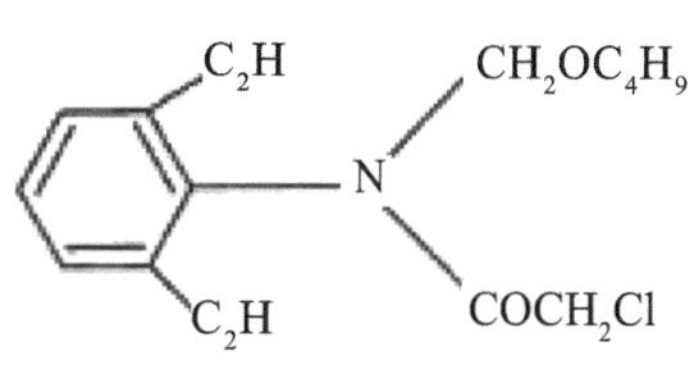

- propanil:3',4'-dichloropropionailide

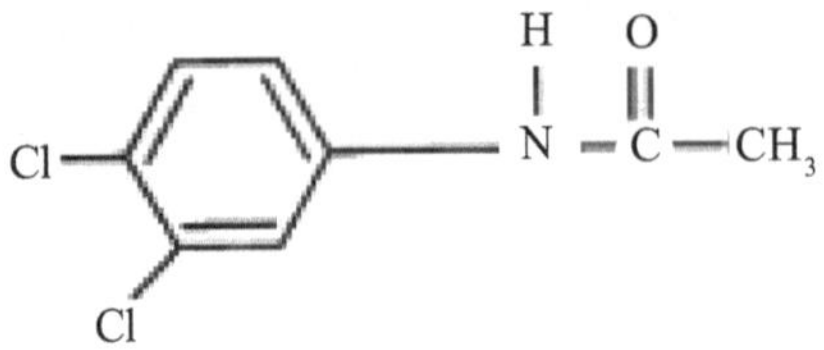

Bipyridiliums

- Diquat.:6,7-dihydrodipyrido (1,2-D,2; 1-c) pyrazinedium ion

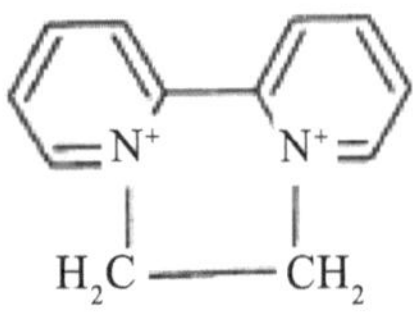

- Paraquat:1,1'-dimethyl-4'bipyridinium ion

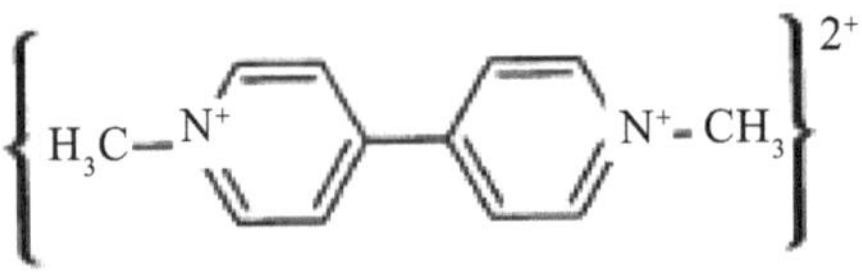

Carbamates

Phenyl Carbmates

- Barban: 4-chloro-2butynyl-m-chloro-carbamate
- Chlro propham:Isopropyl-m-chloro-carbanilate
- Propham:Isopropyl carbanilate

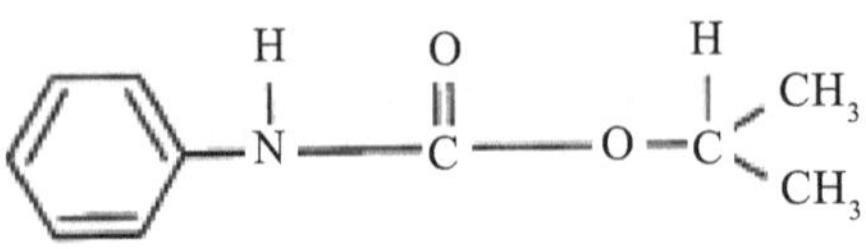

Methyl Carbamates

- Dichlormate:3,4-dichlorbenzyl methyl carbamates

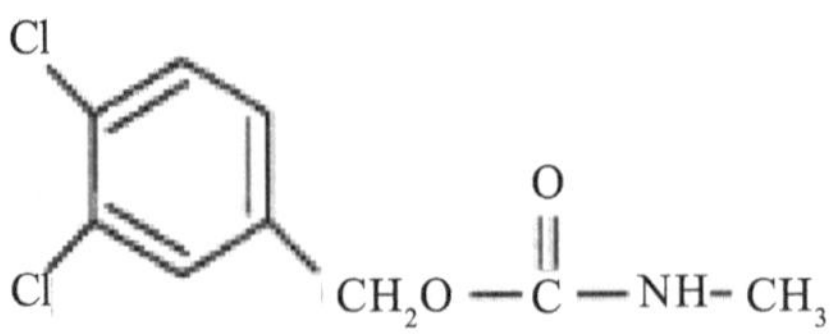

- Dinitramine: N^4,N^4-diethyl-a, a-trifluoro-3, 5-dinitroluene-2, 4-diamine.

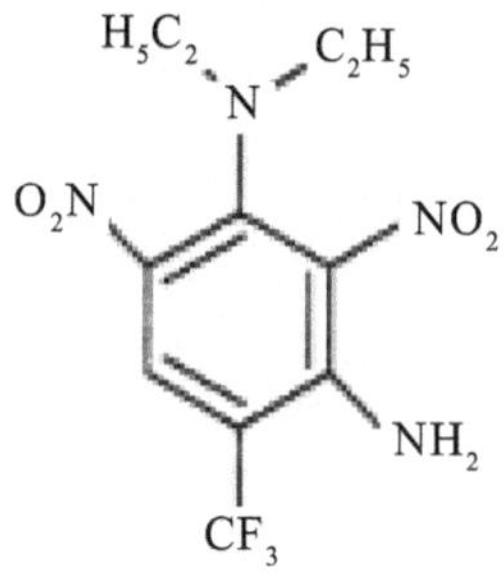

Fluchloralin

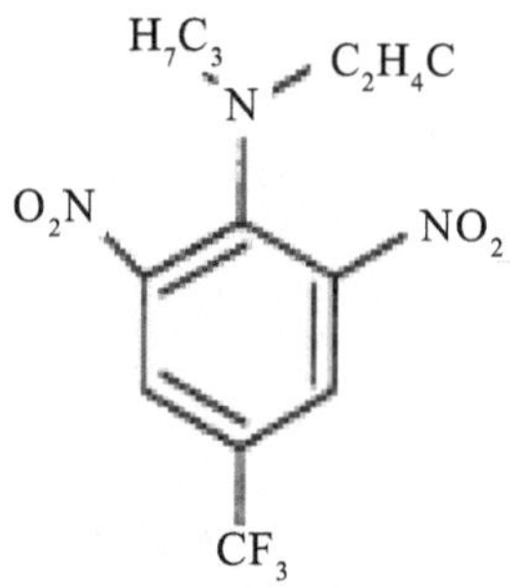

- Trifluralin: α, α, α-trifluoro-2,6-dinitro-N-N-dipropyl-p-toluidine

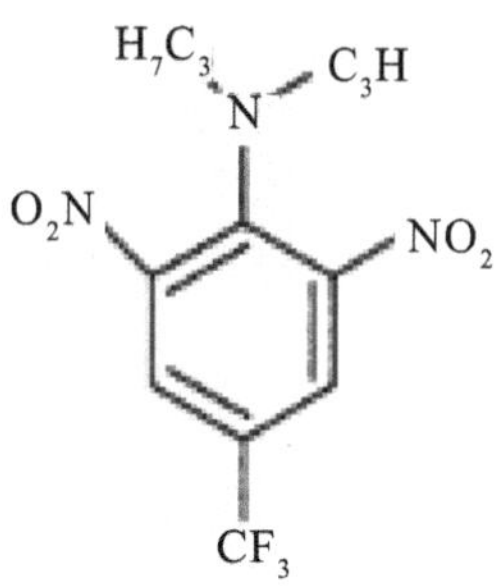

Dipheny Ethers

- Nitrofen:2,4-dichlorophenyl p-nitrophenyl ether

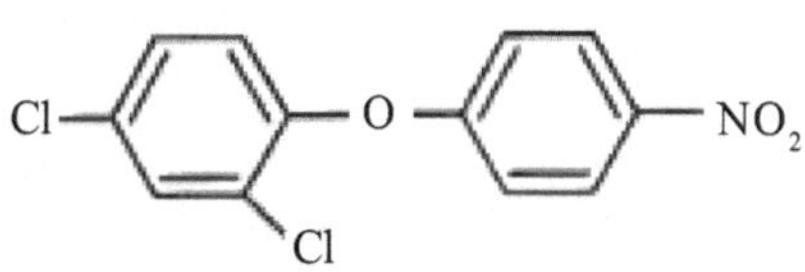

Nitriles

- Bromoxynil. 3,5 dibromo-4-hydroxybenzoitrile

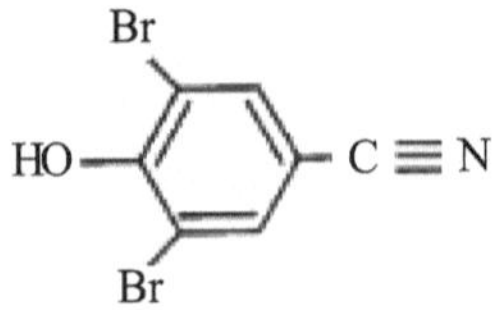

- Ioxynil:4-hydroxy-3,5-di-idobenzonitrile

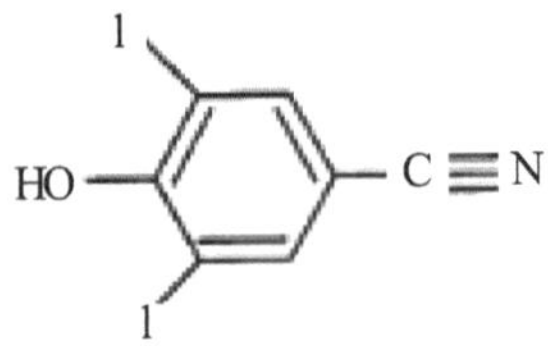

Phenols

- Diniseb:2-sec-butyl-4,6-dinitrophenol

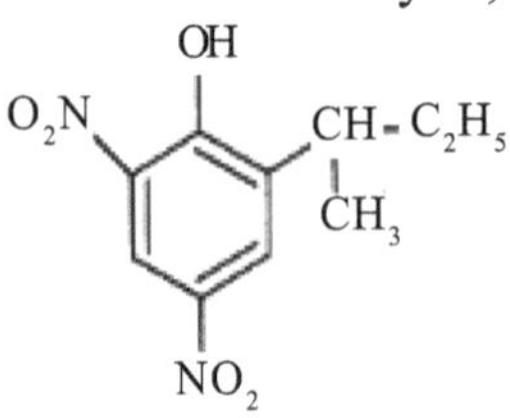

- **PCP:** Penta chlorophenol

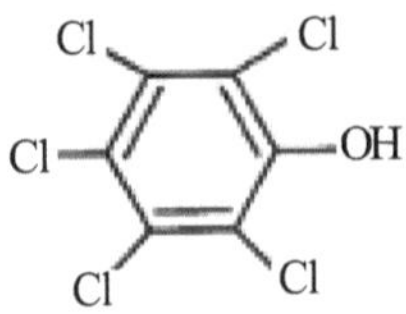

Phenoxy Acids

- 2,4-D:2,4-dichlorophenoxy acetic acid

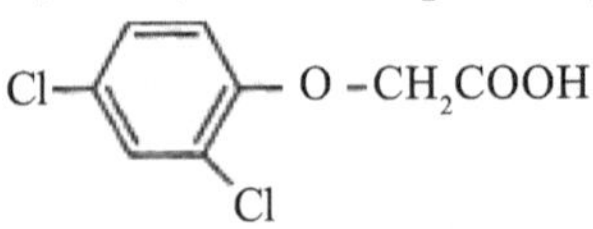

- 2,4-DB: 4(2,4-dichlorophenoxy) butyric acid

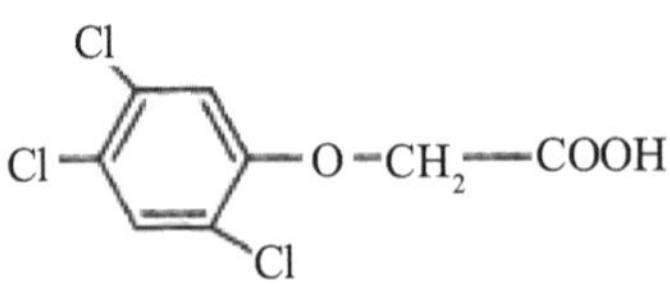

- Diclotop:2[4-(2,4-dichlorophenoxy) phenoxy] prpanoic acid

Thio Carbamates

- Bethiocarb:S-(4-chlorobenzyl) N,N-diethyl thio carbomate
- Diallate: S-(2,3-dichloroallyl) di-iso proppyl thio-carbamte
- Triallate: S-(2,3-3-trichloroallyl) di-iso propyl thio carbamate

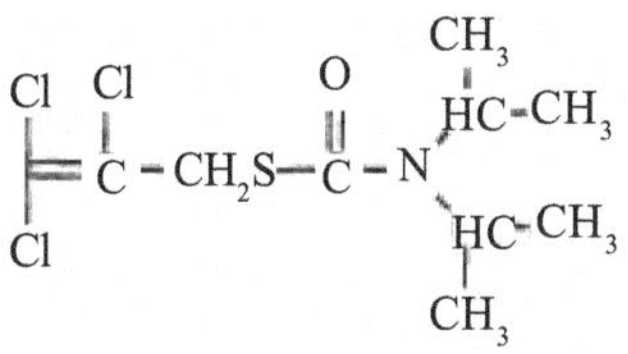

Traiazine

- Atrazine: 2-chloro-4-(ethylamino)-6-(iso propylamine)-S- triazine

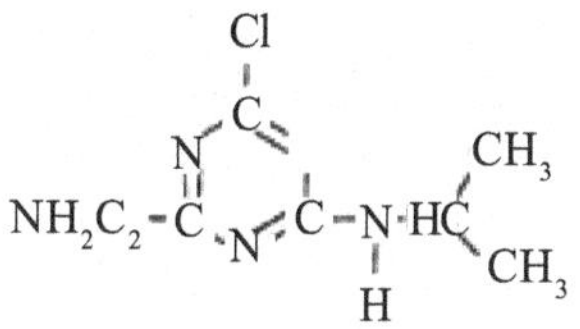

Triazoles

- Amitrole: 3 amino-S-triazole

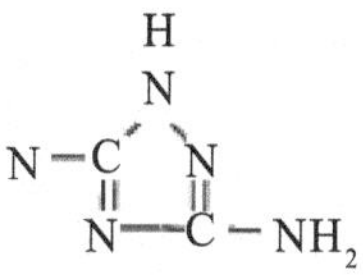

Uracils (Pyridines)

- Bromocil: 5-bromo-3-sec-butyl-6-methyl uracil

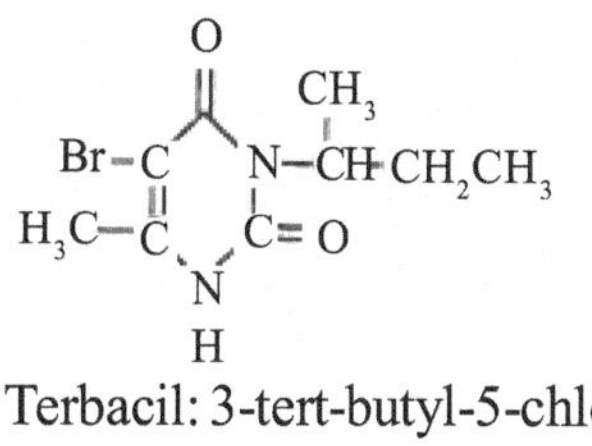

- Terbacil: 3-tert-butyl-5-chloro-6-methylluracil

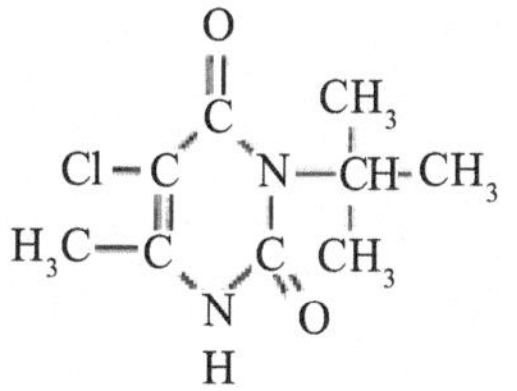

Ureas

Phenyl Ureas Structure

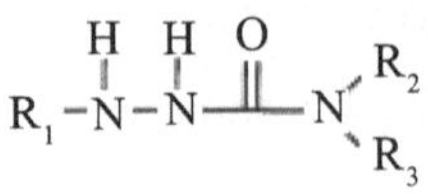

Where, $R_{1=}$Phenyl group

R_2 and $R_{3=}$Alkyl group

- Diuron: 3 (3,4-dichlorophenyl)-1,1-dimethylurea

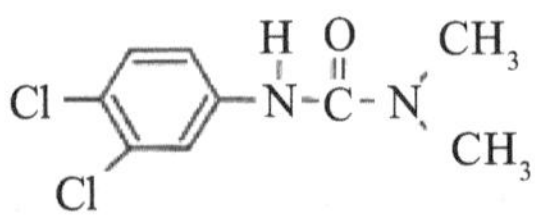

- Fenuron:1,1-dimethyl-3-phenyl urea
- Floumeturon:1,1-dimethyl-3-(a,a,a-trifluoro-tolyl) urea
- Linuron: 3-93,4-dichlorophenyl)-1-methoxy-1-methyl urea
- Methabenzthiauron:3-(2-benzothiazolyl)-1,3-dimethyl urea
- **Benzadex:** (Benzamidoxy) acetic acid

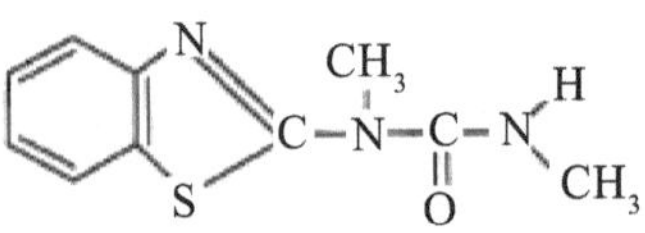

Unclassified Herbicides

- Bentazon: 3-isopropyl-1-1H-,2,1,3-benzothiadiazin-4-(3H)-one 2, 2- dioxide
- Benzadex: (Benzamidoxy) acetic acid
- DCPA: Dimethyl tetrachloro terephthalate
- Glyphosate: N- (Phospono methyl) glycine
- MH (Maleic hydrazide): 1,2-dihydro-3,6-pyridazinedione
- Oxadiazon: 2-tert-butyl-4-(2,4-dichloro-5-isopropxyphenyl)-1,3,4-oxadiazolin-5-one.
- Picloram: 4-amino-3,5,6-trichloropicolinic acid.

Herbicide Activity and Selectivity

Herbicide Activity is related to the phyto-toxic effects of a chemical on plant growth and development.

Herbicide Selectivity is referred as the phenomenon wherein a chemical kills the target plant species in a mixed plant population without harming or only slightly affecting the other plants.

Herbicide activity and selectivity is mediated mechanically, physically, chemically, and metabolically. The various factors that affect herbicide activity and selectivity are as under:

- Chemical and physical properties
- Formulation, time, placement and rate of applicator
- Stage of plant development
- Cultivation practices
- Absorption Translocation
- Physiological differences
- Metabolism
- Environmental factors

Herbicide Formulations

Herbicide formulation is the process by which the technical active ingredients are made ready for use by mixing with liquid or dry diluents by grading/and by the addition of emulsiûers, stabilizers and other formulation adjuvant. It involves conversion of a particular technical ingredient to a commercial product.

1. Active Ingredient

Active ingredient (a.i) is that part of the formulated product mainly responsible for the herbicidal effects. It may be a solid, a liquid or a gas. In general, active ingredients cannot be used for weed control without further processing into a suitable formulation.

During the synthesis of active ingredients formation of small quantities of intermediate byproducts are expected. If they are allowed with the active ingredients even in very small quantities can cause:

- Reduced activity of the active ingredient in the formulation,
- Changes in physical properties of the active ingredient,
- Reduced storage stability of emulsiûable concentrates, and
- Reduction in grind ability of the technical active ingredient.

2. Chemical Stability

Chemical stability of the active ingredient in the formulated product determines its effectiveness. The active ingredient is stated alone and mixed with possible carrier materials for stability before suitable stabilizer is fixed.

Chemical stability depends on reactions between active ingredient and time inert materials used. Generally, formulations containing low concentrations of active ingredient tend to lose active ingredient more rapidly during storage than the more concentrated herbicide forms. Some liquid herbicide products tend to develop gas upon deterioration leading to explosive rupture of containers.

3. Particle Size

Particle size of herbicide formulation inûuences its biological activity and physical properties. Particle size depends upon weed biology, chemical and physical properties of the active ingredients, nature of the target (leaf surface moisture content and soil pH), quality and quantity of the deposit, mode of treatment, etc. Distribution of particle size in a formulation can be determined by microscopic method, sieving elutriation or sedimentation method.

4. Types of Formulations

Herbicides may be solid, liquid or gas. Active ingredients are to be processed: into a suitable formulation for effective weed control.

Wettable Powder (WP)

Wettable Powder contains finely divided solid particles ($<3\mu$) which can be readily suspended in water to form suspension. Wettable powders provide high degree of dispersion of active ingredient for its penetration. A wettable powder, in addition to active ingredient, contains diluents, a surface active agent and an auxiliary material. A wetting agent is also included, if necessary, to increase the retention of the chemical on plant surfaces. Commonly used diluents with low bulk density having high sportive capacity are silica gel, hydrated aluminium oxide, synthetic calcium silicate, etc. Detergents such as sulphonates of alkali metals, sodium d ibutoxy diphenyl sulphonate, sodium alkane sulphonates, etc. act as surface active agents. Sodium salts of sulpho acids and lignin sulpho acids are used as auxiliary materials.

Wettable powder containing 50 to 80 per cent active ingredients are the most widely used. These powders need protection against moisture. They are therefore, packed in polyethylene or aluminium foil liner paper bags or metallic drums. Examples: Atrazine 80% WP, Diuron – 80% WP, Simazine – 50% WP etc.

Soluble Powder (SP)

These herbicides are formulated as water soluble powder and they form a homogenous solution as and when dissolved in water which can be applied to the crop by spraying.

Emulsifiable Concentrates (EC)

The active ingredient (a.i) of this herbicide is dissolved in solvent and an emulsifier is added to it. These materials form stable emulsion, when diluted with water become suitable for spraying on the plant. Examples: Dalapon, Sodium salt of 2,4-D, TCA etc.

Emulsifiable Concentrates upon dilution with water give stable emulsions for spraying plant surfaces. Emulsions at equal concentration of the active ingredient are usually more effective than corresponding Examples: Alachlor, Dialate, Nitrofen, 2,4-D ester etc. suspensions. Usually, two types of EC are made: concentrated emulsion and miscible oils.

Concentrated emulsions are prepared by dispersing an aqueous solution of herbicides in a water immiscible solvent by means of colloidal mill. Only herbicides which are stable to the action of water are suitable for preparation as concentrated emulsion. The miscible oil consists of herbicide, solvent and emulsiûer. When mixed with water, they yield stable emulsions.

Emulsifier act as surface active agents which will readily emulsify the concentrate in the final spray liquid. Calcium sulphonates, rnono esters of sorbitol and monnitol with higher fatty acids, various soaps, etc. can be used as emulsiûer. Emulsion is done, primarily, to disperse the small volume phase (concentrate) in the large volume phase (spray carrier). Each dispersed particle is then completely surrounded by the carrier liquid.

Solution Concentrates (SC)

A solution concentrate consists of a herbicide dissolved in a solvent system to provide a concentrate that is stable in the carrier. Common formulations of this type are water concentrates designed for use in water carriers. Salts of herbicides 2,4-D, MCPA, 2,4,5-TP, sodium salts of pentachlorophenol, etc. are soluble in water. These can be dissolved in water and sprayed. Some types of solution concentrates are oil soluble. Herbicides such as pentachlorophenol, ester formulations of 2,4-D, etc. are soluble in oil. These oil soluble solution concentrates are preparations of various organic solvents.

Soluble Concentrated (WSC)

These herbicides are formulated in the form of soluble liquids which are called water soluble concentrates (WSC). Examples : Amine salt of 2,4-D, 2,4,5 –T, Silvex (2,4,5-TP), Diaquat, Paraquat etc.

Liquid Suspension (LS)

The active ingredient of this herbicide is solubilized in organic solvent which when added to water for spraying forms a liquid suspension. The main drawback of these herbicides is that it requires constant agitation during spraying to avoid settling. Examples : Atrazine, Cyprazine, Nitralin etc.

Granules (GR)

Granules contain 1 to 20 per cent concentrations of herbicides in particular solid carriers suitable for direct ûeld application. In a granular formulation, the technical active ingredient is mixed with or coated on inert carrier material of the approximate particle size of granulated sugar. Granules are generally made, with inert clay. Many materials such as expanded vermiculite, diatomaceous earth, bentonite, illite, limestone, gypsum, sand etc. can be used as carriers. The granules mixed with the solution of toxicant are dried and then packed for subsequent use. Granular formulations can be directly applied to the soil. Examples: Butachlor – G, 2, 4-DEOG etc.

The same herbicide can be made in different formulations. As for examples, 2,4-D when added to sodium, amine and ethyl ester, it forms soluble solid, soluble concentrate and emulsifiable concentrate, respectively.

Fumigants

There are volatile chemicals applied into conûned spaces or into the soil to produce a gas that will destroy weed seeds and act as a soil sterilant. They are often injected as liquids, a few can below soil surface. Fumigants are used for killing deep rooted perennial weeds and for eliminating weed seeds from soil. Herbicides commonly available as fumigants are methyl bromide, metham, allyl alcohol, chloropicrin, and tetra chlorothane.

The herbicides of different formulations such as SP, WP, WSC, EC, etc. are mixed with water, a common carrier of herbicides and sprayed on the soil or foliage. The liquid formulation of herbicides such as EC, WSC etc., can be applied on soil after mixing with sand.

Table 2.4.3: List of important herbicides with trade names and formulation

Chemical Name	Trade Name	Formulation (%)	Source
Alachlor	Lasso	50 EC	Monsanto
Anilofos	Aniloguard, Arozin	30 EC	Aventis
Atrazine	Atrataf, Solaroo	50 WP	Rallis
Butachlor	Machete	50 EC	Monsanto
2,4-D	Knock weed, Fernoxone	Na Salt 80 %	Rhone-poulenec
	Weedmar	Ethyl ester 34 %	Herbicide India Ltd
Chlorimuron	Kloben	20 WP	Dupont
Chlorsulfuron	Glean	75 WP	Dupont
Clodinofop	Topik	15 WP	Syngenta
Clomozone	Command	50 EC	Rallis
Dicofop-methyl	Illoxan	28 EC	Aventis
Diuron	Karmex	80 WP	Dupont
Dalapan	Hexapon	85 WP	BASF
Fluchloralin	Basalin	45 EC	BASF
Fluazifop-butyl	Fusilade	9 EC	Syngenta
Glyphosate	Round-up, Glycel	41 EC	Monsanto/Excel
Glufosinate-ammonium	Basata/Liberty	15 SL	Aventis
Isoproturon	Arelon	50 & 75 WP	Dupont/Gharda
Imazethapyr	Pursuit	10 EC	Cynamide
Linuron	Afalon	50 WP	Aventis
Metolachlor	Dual	50 EC	Syngenta
Metribuzin	Sencor	70 WP	Bayer
Metoxuron	Dosanex	80 WP	SeariInd Ltd
Metsulfuron-methyl	Algrip	20 WP	Dupont
Oxadiazon	Ronstar	25 EC	Rhone-Poulenc
Oxyfluorfen	Goal	23.5 EC	Rohm & Hass
Paraquat	Gramaxone	24 EC	Syngenta
Pendimethalin	Stomp	30 EC	Cynamid
Pretilachlor	Rifit	50 EC	Syngenta
Pretilachlor + safener	Sofit	50 EC	Syngenta
Sulfosulfuron	Leader	75 WP	Monsanto
Thiobencarb	Saturn	50 EC	Pesticide Ind Ltd.
Trifluralin	Treflan	48 EC	De-nocil/Gharda

Effectiveness of Herbicides

It depends upon the following factors:

- Time of herbicide application: Whether applied as pre-plant or pre-emergence or post emergence
- Type of weed
- Climatic conditions
- Concentration of spray material
- Formulation of herbicides
- Type of carrier used in herbicide formulation
- Use of surfactants
- Soil type
- Type of sprayers and nozzles
- Previous history of weeds
- Herbicide mixtures

Adjuvants and Safeners

Certain chemicals are used for improving the herbicidal effects, selectivity and to prolong self life of the concentrate. Some substances are also used along with herbicides to protect crop plants from possible damage by the herbicide.

Adjuvants

Adjuvants are chemicals used to improve the herbicidal effects for efficient weed control. Adjuvants may not increase the innate activity of any herbicide, but they improve its availability at the action site in plants. Adjuvants are also added to herbicide formulations to improve herbicide selectivity to the non target plants, to render herbicide safer for the user and to prolong self life of the concentrate. There are eight kinds of adjuvants.

Surfactants

Surfactant is a material which favours or improves the emulsifying, dispersing, spreading, wetting or other surface modifying properties of herbicide formulation. Surfactants aid in wetting the waxy leaf surface with aqueous herbicide sprays (**Wetting agents**), in spreading the hydrophilic herbicides uniformly over the foliage (**spreaders)** and in the penetration of the herbicide in to the target sites (**penetrants**). Examples of surfactants are:

Nonionic: S- 145 or Tween – 20 (poly oxy ethylene sorbitan mono laurate).

Surfactant WK (Dodecyl ether of poly oxy ethylene glycel.)

Anionic: Santomerse (dodecyl benzene sodium sulphate) Vatsol – OT (sodium di octyl sulpho succinate).

Cationic: Quaternary – O (alkylimidozolinium chloride) Aliquat – 4 (lawryl trimethyl ammonium chloride).

Stabilisers

Stabilisers are chemicals used for obtaining stable emulsion and suspension in the spray tank individually, known as emulsifiers and dispersants respectively. Emulsifier causes an emulsion concentrate to disperse spontaneously in to small stable droplets when added to water. It substitutes for constant physical agitation of spray liquid during operation. Examples of emulsifiers are 15-S-particles of wettable powders in suspension in water even after initial vigorous agitation has been withdrawn from the spray tank. Multi film, tryad and bio film are examples of dispersants.

Coupling Agents

Coupling agent (solvent and co-solvent) is a chemical used to stabilize herbicide in concentrated form such that the resulting solution is soluble or miscible with water in all proportions. Acid form of 2,4-D which is almost insoluble in water can be dissolved in polyethylene glycol to make it water soluble. Commonly used solvents are lanolin, carbo wax, benzene, xylene, carbon tetrachloride, methyl chloride, acetone, etc.

Humicants

Humicants such as glycol prevent rapid drying of herbicide sprays on the foliage thus providing an extended opportunity time for herbicide absorption by the plant. They are also called as **hygroscopic agents**, since they are hygroscopic in nature.

Deposit Builders

Deposit builders (stickers or filming agents) are chemicals added to herbicide concentrates to keep the toxicant in intimate contact with plant surface. They minimize wash off of the toxicant from the foliage due to rain. Several petroleum oils and stickers act as deposit builders.

Compatibility Agents

Fertilizers, pesticides and herbicides in liquid form can be applied in a single spray to save application cost. Compatibility agents like complex are used to intimately mix fertilizers and pesticides in spray liquids/

Activators

Activators are chemicals which enhance the phytotoxicity of herbicides. Phyto bland oils increase the foliar activity of atrazine against annual grasses as they act as strong penetrants. Iso paraffinic oil used as diluent accelerates atrazine and dinoseb action against *Imomo eahederaceae*. Ammonium thocyanate, an activator of amitrole, enhances the amitrole mobility leading to effectiveness of amitrole. Nitrogen fertilizers like urea, ammonium chloride and ammonium nitrate are used to improve 2, 4-D phyto toxicity for minimizing its rate of application.

Drift Control Agents

Invert emulsions are used to increase the size of herbicide spray droplets for minimizing serious hazards to non target plants. Certain chemicals can also be used to increase the droplet size of sprays. Thickening agents with large organic compounds in aqueous solution behave like gels and make the spray liquid viscous. Such viscous sprays behave like invert emulsions and produce large sized spray droplets. Sodium alginate and decagin act as thickening agents. Particulating agent comprising of lightly cross linked swellabel polymer imbibe liquids (water and oil) to form particulate aqueous and particulate oil as the case may be. The drift potential from particulate sprays is less than for sprays of thickened or non thickened liquids. Norbak is an commonly used particulating agent.

Safeners

Herbicide safeners, protectants or **antidotes** are substance used along with herbicides to product crop plants from possible damage by the herbicide. Naphthalic anhydride (NA) or dichlormid can protect maize, sorghum and rice injury by clorsulfuron. Use of NA also protect crops form damage by imazaquin. Several safeners can be formulated as seed dressings. NA protects rice against alachlor. Similarly, Cyometrinil protects sorghum against metalochlor.

Safeners are, generally, most effective when mixed with herbicide and applied to the soil. The protection is more marked with maize, sorghum and rice. There has been limited success in protecting broadleaved crops. List of safeners introduced in agricultural practice is given in Table.

Safeners Introduced Into Agricultural Practice (Sankaran *et al.*, 1993)

S.No.	Common name	Chemical name	Appln. Method	Protected plant
1.	NA	Naphthalene-1,8-dicarboxlicacid anhydride	SD	Maize
2.	CGA-133205	0-(1-3-dioxolan –2-yl methyl) 2,2,2-trifluoro-4-chloroacetophenone oxime	SE	Sorghum
3.	Fenclorim	4,6-dichloro – 2 phenyl-pyrimidine	TM	Wheat
4.	Dimepiperate	N-(1,1-dimethylbenzyl-thio-thiocarbonyl) –pi[erodone	TM	Rice

Herbicide Absorption and Translocation

Absorption : It is the movement of a herbicide from the surface into the plant body in a limited zone.

Translocation: It is the transfer of a herbicide, food or other materials from one part of the plant to another.

Absorption- Pathways in Foliar Absorption

The absorption of herbicides involves their entry from plant surfaces into the cytoplasm of epidermal cells of shoots or roots. In this process a herbicide may come across five possible barriers. These are (i) surface waxes and hair (ii) cuticle (iii) periderm (iv) cell wall and (v) plasma lemma.

Plant surfaces exposed to full sunlight developed thicker wax coatings than those exposed to low lux light. Waxes and hairs are lipo philoc in nature (Molecules or surfaces that are attracted by oil-like surfaces and molecules, but are repelled by water) and they interfere in wetting of leaf surfaces with lipo phobic sprays (lipo photoic = hydrophilic) (molecules or surfaces that are attracted towards water such as sodium chemide. They are repelled by oil like surfaces or substance).

The relative thickness and composition of cuticle on plant foliage determine in nature of leaf surfaces and consequently their wettability with the herbicide spray liquids cuticle carries both hytrophilic (-OH, -COOH) and lipophilic (-CH2,-CH3) groups which import to it a hemi hyotrophilic character. In general young leaves and those growing in shade are much less cutimsed than mature leaves of those growing continuously in bright sun respectively. The roots of plants have no cuticle, therefore, they absorb hyotrophilic substances more rapidly than the lipophilic components. Hydrophilic substances find easier entry thro' cell walls than the lipophilic substances. The mole culourstruchires and formulations of herbicides are very important in determining in foliar absorption of herbicides. In summary, the non polar compounds are absorbed more readily by the foliage than the plan compts.

Note: Polar compounds: Polar compounds have molecules with electropositive and electro-negative regions and they tend to ionize Eg. $Na^+ Cl^-$,H^+2o^-.

Non-Polar compounds: Molecules with no surface distribution of charges are sand to be apolar **Eg**. Hexane, Cyclohexane

Absorption of Herbicides Form Soil

The absorption of herbicides applied to soil is usually associated with roots of the plants herbicides may also he absorbed by the swelling seeds and coleoptile, epicotyl and hypocotyls of the emerging seedlings. The shoot zone of germinating plants in soil (Colloptile etc.) has been revealed as major site of absorption of many soil applied herbicides.

Eg. EPTC, Trifluralim Effectively absorbed when applied to the subsurface, shoot zone of the plants, 2,4-D, Simazine are mainly absorbed from the root zone. Herbicides are also absorbed by rhizomes, stolons and tubers of perennial plants. Roots being devoid of culticle, permit easier entry of polar compounds than the non-polar compounds.

Mechanism of Herbicide Absorption

1. **Absorption thro' cuticle and cell walls**

   Hydrophilic ions and molecules enter plants by the process of diffusion whereas the lipophilic substances enter by their lipid solubing action. The lipophilic spray liquids solubilize the nature lipid molecules of cuticle and epidermal walls and increase their permeability to the herbicide. Many Adjuvants import certain lipophilic properties to the aqueous spray liquids and aid entry of hydrophilic substances thro' the cuticle and cell walls.

2. **Absorption thro' plasma membranes**

   The plasma membranes comprising plasma lemma and tono plast are made up chiefly of lipids and proteins and are lipophilic nature. The lipophilic herbicides as herbicidal oils and oil emulsions pass thro' them easily at rates roughly proportional to their fat solubility and they reach the epi demeal cytoplasm rapidly.

After a herbicide has been absorbed by the epidermal cells of plants, it moves across the thickness of the root or shoot, as the case may he , primarily in ionic form till it reaches close to vascular vessels – the Phloem and xylem.

The movement of herbicides may occur either thro' apoplast or thro Symplast. In both cases, the movements is a process of diffusion although in the symplast, cytoplasmic streaming and active accumulation along the electrochemical potential gradient may also help the movement of herbicides.

Apoplast : Inter connecting cell walls and intercellular spaces, including the water (or air) filled xylem elements – a non living system.

Symplast : The system of interconnected protoplasm excluding vacuoles – a living system.

Translocation of Herbicides

Once a herbicide is absorbed into the plant system, it moves either apoplastically or symplastically. The term apoplast is referred to the system of the non living inter connecting cell walls, intercellular spaces and the water filled and air filled xylem elements. Thus, apoplastic movement takes place predominantly thro' the xylem and intercellular spaces. The word symplast is referred to the system of inter connected protoplasm that is connected from cell to cell by means of plasma desmata, excluding the vacuoles. Generally, the symplast includes the phloem transportation mechanism and the apoplast includes the xylem stream.

Translocation of Soil-Applied Herbicides

The soil applied herbicides taken up by the root or shoot penetrate into the xylem and move upward apo plastically in the transportation stream. Even highly toxic herbicides can be absorbed from the soil and readily translocated to all parts of the plant without injuring the xylem which is principally a non living tissue.

Triazine, vracil and urea herbicides which have very low ability to enter the symplast in leaves, move acropetally via apoplast to the leaf meso phyle very rapidly with water. They are translocated to the leaves, stems, and buds as well.

Translocation of Foliage Applied Herbicides

The two main transport pathways are the phloem and the xylem. The herbicides move along with the assimilates from the leaves in a stream through the phloem utilizing the source to sink principle, the main sources being the expanded and growing leaves and the sinks being the shoot tips, buds, roots, fruits and other plant parts undergoing development and expansion. A strong sink is essential for an active translocation of herbicide from leaves. The direction of movement is from the regions of synthesis of food or hydrolysis of food reserves to regions of food utilization.

Herbicides like picloram and dolapon are translocated thro' the xylem when absorbed by roots and thro' phloem when absorbed by leaves. A herbicide which is translocated primarly thro' the phloem can also 'leak out' of the symplast and enter the xylem circulation.

Translocation Through Phloem

The phloem mobile herbicides move along with photosynthates in the phloem tubes of the several mechanisms of translocation and along with it the herbicides, thro' phloem tubes, the mass flow hypothesis is widely accepted since it is applicable to acropetal, basipetal as well as in directional translocation of food material and herbicides in plants. It also explains well the high rates of translocation of certain herbicides observed in plants; for instance, under favourable conditions 2,4-D may translocate through the phloem as fast as 160 cm per hour. Herbicides can translocate through only healthy phloem tubes that are in state of a active food translocation when for some reason the phloem tubes are damaged, even a normally translocated herbicide will fail to translocate in plants. Dalapon when it is applied at low conc. of 0.2-0.5% to a healthy grass clump, it translocates easily to its roots. But when the spray conc is high (1to 2%), dalapon will quickly brown the treated foliage without translocating apprecially to the grass roots. Likewise paraquat and diquat herbicides are immobile in plants mainly because of their rapid phyto toxicity to leaf tissues even at low concentration. The low phloem mobility of 2,4-D in plants is also ascribed to disruption of phloem tubes by proliferation of adjoining cells by the herbicide.

Two important characteristics of translocation of herbicides thro' phloem are:

i) A herbicide translocates in plants along with food manufacture (ie mature leaves to the site of food consumption or storage (viz., growing points, young leaves, roots and rhizomes). The site of food synthesis or food export in plants is called 'source' and the site of its ultimate lodgement, the 'sink'.

ii) During translocation thro' phloem, the herbicides by-pass mature leaves because such leaves synthesize and export food but do not import it. The phloem mobile herbicides enter only the activity food importing organs of plants.

Translocation Thro' xylem

Root mobile herbicides translocate acropetally along the transpiration stream. Reaching all live leaves, both mature and immature, above the point of application of the herbicide. A major difference in the translocation behaviour of a phloem mobile and a xylem mobile herbcide is that the later (xylem) does not by-pass mature leaves on its way to the plant tops, whereas the former (phloem) does so. Also, a xylem mobile herbicide translocates only in a direction acropetal to the point of its application while phloem mobile herbicide translocates both in acropetal and basipetal directions. But a herbicide like pidoram when applied to an intact root, translocates acropetally up to its crown level and then it moves into the phloem. Once the picloram has reached the phloem tubes, it can translocate both acropetally and basipetally to its sinks by-passing the mature leaves.

When a xylem mobile herbicide is placed on a plant shoot, it moves thro' the apoplast and is translcoated upward in the xylem. It will not show any basipetal translocation unless it was capable of leaking into the adjoining phloem tubes. Foliar xylem mobile, will bill only shoots of established annual grasses, leaving their crown buds, roots and hidden meristems unhurt. This is because these herbicides do not translocate form the treated leaves, but cause contact injury to the foliage. On the other hand, when the two compounds are applied to the soil, these are absorbed by the grass roots and are translocated through xylem to the meristematic tissues, crown buds and blades which may be damaged seriously.

Common Types of Translocated Herbicides

The translocation behaviour of herbicides differs with the herbicide structure, and sometimes it may alter with the plant species, environment etc. depending upon these faiths, five important types of translocated herbicides are recognized

1. **Shoot immobile herbicides**: These are foliar applied herbicides that translocate only in the acropetal direction thro' xylem. They do not translocate from leaves to the roots of the treated plants. Eg: Atrazine and diuron
2. **Shoot mobile herbicides:** These are foliar applied herbicides that show both acropetal and basipetal translocation through phloem, sometimes simultaneously. They translocate both to shoots and roots of the treated plants. **Eg.** Amitrole, Glyphosate, Picloram, dalapon, dicambo.
3. **Circulatory herbicides:** These are foliar applied herbicides capable of frequent phloem-xylem interchange. **Eg.** MH,, 2,3-6 TBA, picloram, dalapom
4. **Reverse xylem mobile herbicides:** These are foliar applied herbicides that translocate to roots thro' xylem in acute water deficiency conditions. **Eg.** arsenic trioxide and Na cherate. Paraquat also exhibit reverse flow in xylem when it is applied to leaf in subdued light.
5. **Root mobile herbicides:** These are root applied herbicides that translocate freely in xylem tissues to reach the plant tops **Eg:** dicamba, tricamba, uracils, S-triazines and substituted ureas.

2.5

Herbicide Application Techniques

Herbicide Application

Generally, herbicides are applied in the form of solution or granules. Of the two, spraying solutions of herbicides formulated as wettable powders, soluble powders and emulsifiable and solution concentrates is more common. When herbicides are formulated as granular or encapsulated materials, they are applied by hand or with the help of a granular applicator.

Soil application	Foliar application
a) Surface	i) Blanket spray
b) Sub surface	ii) Directed spray
c) Band	iii) Protected spray
d) Fumigation	iv) Spot treatment
e) Herbigation	

Herbicide Spraying

A spray is defined as a liquid discharged into particles and scattered as dispersed droplets. Herbicides are mixed with a suitable carrier to facilitate distribution and even coverage over the area to be treated. Water is the most common carrier for herbicide spraying. In a spray solution, the herbicide is dissolved, emulsified or suspended in water. Herbicide application is directed to the soil in the case of (PPL) pre-planting and pre-emergence (PRE) herbicides and to the foliage in the case of post-emergence (POST) herbicides.

Spraying Equipment

The basic components of the sprayer include the tank to hold the spray solution, the nozzle through which the spray solution is discharged, the pump for generating the necessary pressure, a filter or strainer, a pressure regulator and a hose.

Methods of Application of Herbicides

Spraying and Broadcasting are the major methods by which herbicides are applied to different crops.

Factors Influencing the Methods of Application are

Weed-crop situation, Type of herbicides, Mode of action and selectivity, Environmental factors, Cost and convenience of application

Depending on the Target Site, the Herbicides are Classified in to

i) Soil applied herbicides

ii) Foliage applied or foliar herbicides

Different methods by which these herbicides are applied is tabulated below

Soil Application of Herbicides

a) **Surface application:** Soil active herbicides are applied uniformly on the surface of the soil either by spraying or by broadcasting. The applied herbicides are either left undisturbed or incorporated in to the soil. Incorporation is done to prevent the volatilization and photodecomposition of the herbicides.

E.g. Flucholralin - Left undisturbed under irrigated condition

- Incorporated under rainfed condition

b) **Subsurface application:** Application of herbicides in a concentrated band, about 7-10 cm below the soil surface for controlling perennial weeds. For this special type of nozzle is introduced below the soil under the cover of a sweep hood.E.g. Carbamate herbicides to control *Cyperus rotundus,* Nitralin herbicides to control *Convolvulus arvensis.*

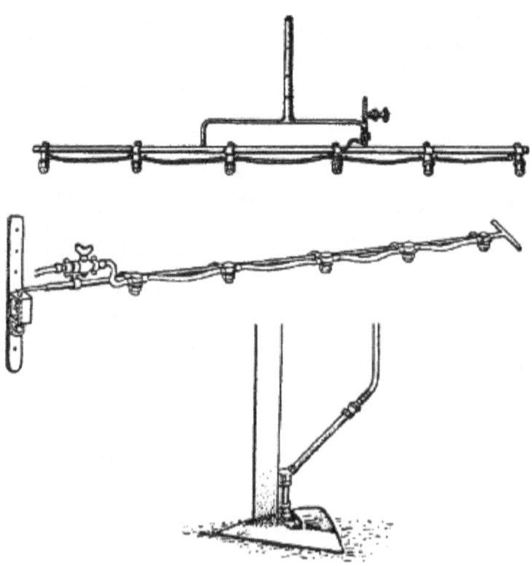

Fig. 2.5.1: Top and middle =Two kinds of tractor –mounted multi-nozzle spray booms. Bottom=A spray nozzle fitted to a sweep for sub-surface application of a herbicide.

c) **Band application:** Herbicides are applied to a restricted band along the crop rows leaving an untreated band in the middle of inter-rows. Later on these left out inter rows are cultivated to remove the weeds. Saving in cost is possible here. For example, when a 30 cm wide band of a herbicide applied over a crop rows that were spaced 90 cm apart, then two-third cost is saved.

d) **Fumigation:** Application of volatile chemicals into confined spaces or in to the soil to produce gas that will destroy weed seeds is called fumigation. Herbicides used for fumigation are called as fumigants. These are good for killing perennial weeds and as well for eliminating weed seeds.

e.g. Methyl bromide, Metham

f) **Herbigation:** Herbicides are applied with irrigation water both by surface and sprinkler systems. In India farmers apply Fluchloralin for chillies and tomato, while in western countries application of EPTC with sprinkler irrigation water is very common in Lucerne.

Foliar Application

i) **Blanket spray:** Herbicides are applied uniformly to standing crops without considering the location of the crop. Only highly selective herbicides are used here. E.g. Spraying 2, 4-Ethyl Ester to rice three weeks after transplanting

ii) **Directed spray:** Application of herbicides on weeds in between rows of crops by directing the spray only on weeds avoiding the crop. This could be possible by use of protective shield or hood. For example, spraying Glyphosate in between rows of tapioca or sugarcane using hood to control *Cyperus rotundus*.

iii) **Protected spray:** Applying non-selective herbicides on weeds by covering the crops which are wide spaced with polyethylene covers etc. This is expensive and laborious. However, farmers are using this technique for spraying Glyphosate to control weeds in jasmine, cassava, banana.

iv) **Spot treatment:** It is usually done on small areas having serious weed infestation to kill it and to prevent its spread. Rope wick applicator and Herbicide glove are useful here.

Computation of Herbicide Dosage

In designing a chemical weed control programme, working out the final commercial product required for application is an important step. If correct dose is not worked out based on the recommendation for application, then there are chances that you may end with either no control or sometimes you may loose your crop and even chances are there that your crop in rotation may get affected.

The herbicides are either recommended on active ingredient basis or on formulation basis. For example Butachlor 1.25kg a.i./ha as pre-emergence is the recommendation on active ingredient basis and the same can be recommended as Butachlor 50% EC at 2.50 lit on product basis if the commercial product available has 50% active ingredient in it.

Calculation of actual amount of final product based on recommendation of active ingredient/ha

Formula

Amount of commercial product/ha

$$= \frac{\text{Rate of active ingredient/ha X 100}}{\text{\% active ingredient in commercial formulation}} \text{X Area}$$

Quantity of herbicide formulation required for a given area

F= R × 100/C × 10000) × A

F= Formulated product required (kg or L/ha)

R= Rate of a.i. kg/ha to be aprayed

C=% concentration of a.i. in the formulation

A= Area to be sprayed, m^2

Spray Calibration

Accurate calibration of spray equipment is essential if herbicides are to be used safely and effectively. A sub lethal dose of spray will fail to give satisfactory control of weeds. An overdose will increase costs besides killing a crop or resulting in accumulation of toxic residues in the soil. Achievement of a desired application rate is dependent on selection of the right combination of nozzle type, size of nozzle orifice, spraying pressure and spraying speed. The best way to calibrate is to spray an area of known size and measure the volume of spray delivered from the tank. This may be done using different nozzles at different spray pressure and walking speeds. Care must be taken to see that speed and pressure do not vary significantly from the calibration test to actual field spraying.

The process to find out the exact quantity of carrier (water) required for spraying a given area under a set of conditions is called "calibration". Calibration of a sprayer is a must for spraying herbicide in the field. Calibration can be done by filling a known quantity of water into the sprayer and spraying it over a measured area.

The quantity of the water required for spraying a given area can be calculated by using the following formula.

$$Q = \frac{(V1\text{-}V2)}{a} \times A$$

Where,

Q = Quantity of water required for area to be sprayed (liters)

V_1 = Quantity of water taken initially

V_2 = Quantity of water remained after spraying

A = Area to be sprayed m^2

a = measured area sprayed m^2

The Area Covered Per Hour can be Calculated as Shown Under

Area (ha)/hr

$$= \frac{\text{Walking speed (km/hr) x m/km x spray width (m)}}{\text{10000 (i.e m2/ha)}}$$

For example if a person is walking at 1km/hr covering a swath of 0.6 m wide, the area covered/hr is

$$\frac{1 \text{ x } 1000 \text{ x } 0.6}{10000} = 0.06 \text{ ha/hr}$$

At this rate, it will take 16 hr 40 minutes to cover an area of one hectare. At a spray discharge rate of 30 litres/hr, it would require a spray volume of 500 litres /ha. The rate of spray applied can be determined by the following equation.

$$\text{L ha}^{-1} = \frac{(\text{L min-1 }) \text{ x } 6{,}000}{(\text{km ha-1 }) \text{ xW}}$$

Where,

l min^{-1} : discharge per nozzle, in litres per minute

km ha^{-1} : Walking or ground speed in kilometres per hour

W : Spray width in centimetres

6,000 : a constant to convert litres per minute, kilometres per hour, and centimetres to litres per hectare

If more than one nozzle is used, spraying is done in bands or in crop rows. Then, spray width (W) can be calibrated as shown below.

Row spacing (or band width)

No. of nozzles per row (or band)

$$W = \frac{\text{Row spacing (or band width)}}{\text{No. of nozzles per row (or band)}}$$

Herbicide Application Equipments

Spraying Equipment

Herbicides are generally sprayed using hand-operated sprayers. Depending upon situations, power operated sprayers are also used. Based on the volume of spray liquid used the sprayers are classified into

1. High volume sprayers: 500-750 lit. per hectare E.g. Knapsack / Bakpak sprayer
2. Medium volume sprayer: 250-300 lit. per hectare E.g.Rocker sprayer
3. Low volume sprayer: 150-200 lit. per hectare E.g. Motorised (power) sprayer (Aspee)
4. Ultra low volume sprayer: Less than 5 lit.per hectare e.g. Controlled droplet applicator (CDA) – micron herbi

High volume sprayers are mostly used for herbicide applications because they avoid drift hazard. Based on the mode of operation, sprayers are classified into two types;

i) Hand operated sprayers

   e.g: Knapsack sprayer, Bakpak sprayer, Rocker sprayer, *etc.*

ii) Power driven sprayers

   e.g: Power sprayers, power tiller drawn, tractor drawn sprayers, *etc.*

Basic Components of a Sprayer

- A tank to hold the spray solution
- The nozzle through which the spray solution is discharged
- The pump for generating the necessary pressure
- A filter or strainer
- A hose with lance or boom to hold the nozzle and for directing the spray

Tanks

Tanks are constructed of brass, galvanized iron, stainless steel, fibreglass or glass reinforced plastic (FRP). Of these, the stainless steel and FRP tanks are resistant to corrosion by all types of agricultural chemicals. Tanks with resin-bound glass fibres or lined with inert plastic materials resist corrosion. Tanks vary in capacity, ranging from one litre, or even less in the case of a hand-atomizing sprayer, to several hundred litres in the case of power-driven tractor-mounted sprayers. The most commonly used hand operated knapsack and bakpak sprayers have a tank capacity ranging from 10-15 litres sufficient to discharge in 15 to 30 min. depending on the nozzle type. All tanks should have a large inlet to allow easy loading and periodic inspection of the inside.

Sprayers have agitators to prevent separation and settling of emulsions or suspensions. There are two types of agitation: mechanical and hydraulic. Mechanical agitation is achieved by a series of paddles on a shaft that runs horizontally through the tank or by means of a propeller at one end of the tank. The paddles or propeller are power-driven at a low speed. Mechanical agitation is generally used for emulsions with a high percentage of oil and for wettable powders. Hydraulic agitation is achieved by routing some of the pressurized spray liquid back into the spray tank. Hydraulic agitation is most often used for readily soluble or self-emulsifying materials. Generally mechanical agitation is more efficient in terms of power requirement and uniformity in dispersion.

Nozzles

The nozzle is the most important component of the sprayer, as it inûuences uniformity in spraying, rate of herbicide application, spray drift, spray pattern and droplet size. It breaks the pressurized spray liquid into droplets or application to the target. The function of the nozzle is to accelerate and disintegrate the flow of fluid spray passing through it into droplets to form a spray.

A nozzle consists of a body, cap, lip or disc, and core. In some cases the nozzle body is incorporated with a lance or boom. A nozzle filter is sometimes used with the nozzle to prevent blocking of the nozzle orifice by dust and other foreign particles. Different types of nozzles are available to produce varying spray patterns and droplet sizes. In the cone type nozzle, the liquid is forced through one or more tangential or helical passages into a swirl chamber through which the liquid passes to a circular orifice at a high rotary velocity to form an air core within the orifice and swirl chamber. The liquid emerges from the orifice as a hollow cone. In the case of a solid cone nozzle, the liquid also passes centrally through the nozzle to fill the air core. Different discharge rates and cone angles can be obtained by changing the disc and core. The hollow cone nozzles produce the heaviest droplet distribution on the edges of the pattern. Cone nozzles are widely used for insecticide and fungicide applications as also for spot application of POST herbicides.

Herbicide spraying is mostly done with fan and flooding or deflector type nozzles as they produce a more even distribution of spray and uniform coverage than the cone-spray types. In the case of a fan nozzle, the tip has a lenticular or rectangular orifice behind which two streams of liquid meet because of the shape of the bore. Fan nozzles are ideal for spraying flat surfaces and for band spraying in row crops. In the flooding nozzle, also known as deflector, anvil or impact nozzle, a jet of liquid passes through a relatively large orifice and strikes a smooth surface at a high angle of incidence to form a fan-shaped spray pattern. The flooding-type nozzles deliver coarse droplets of fluid spray under pressure, thus minimizing drift. They also produce a wide-angle fan pattern of spray. As the nozzle orifices are large, clogging is considerably reduced.

Types of nozzles: Nozzles vary according to size and spray pattern developed by it. They are Cone (hollow & solid), Fan and deflector (flood jet).

Cone Type Nozzle

In cone type nozzle, (Fig.2.5.2) the liquid is forced through one or more tangential or helical passages into a swirl chamber through which the liquid passes to a circular orifice at a high rotational velocity to form an air chore within the orifice and swirl chamber. The liquid emerges from the orifice as a hollow cone or solid cone depending upon the disc or core of the nozzles. Cone nozzles are widely used for insecticide and fungicide applications and sometimes for spot application of post emergence herbicides.

Solid Cone (Image) Solid Cone (Line Diagram) Hollow cone (Line diagram)

Fig. 2.5.2: Solid cone (image), Solid cone (Line diagram) and Hollow cone (Line diagrams)

Fan Type Nozzle

Herbicide spraying is mostly done with fan and flooding or deflector type nozzles (Fig.2.5.3) as they produce a more even distribution of spray and uniform coverage than the cone types. Fan nozzles are ideal for spraying herbicides on flat surfaces and for band spraying in row crops. The tip of fan nozzle has a lenticular or rectangular orifice behind which two streams of liquid meet and because of the shape of the bore it develops a fan type pattern.

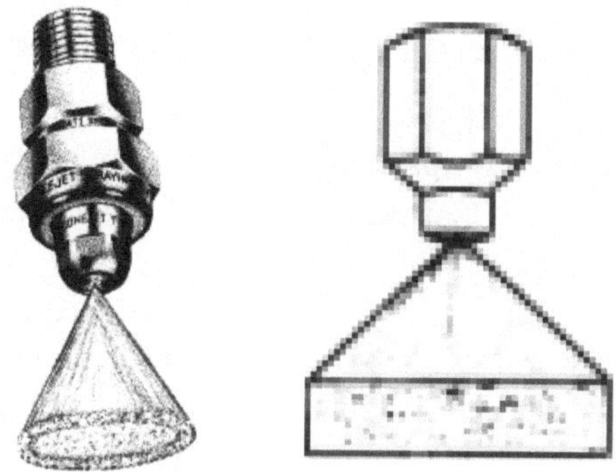

Flat fan (Image) Flat fan (Line diagram)

Fig. 2.5.3: Fan type nozzles

Flooding or Deflector Type Nozzle

It is also known as deflector, anvil or impact nozzle. A jet of liquid passes through a relatively large orifice and strikes a smooth surface at a high angle of incidence to form a fan shaped pattern. The different types of flood jet nozzle are available. Their discharge and spray angle under different pressure levels are given below.

For best results, the hand operated knapsack sprayer should maintain a pressure of 10 to 15 pounds per square inch (0.7 to 1.05 kg/cm^2). The most suited flood jet nozzle for herbicide spraying is WFN 62.

Fig. 2.5.4: Floodjet deflector

Pumps

Pumps are devices that deliver fluids under pressure. The pressure provides the energy that permits the spray liquid to be broken into particles by the nozzles and determines the rate of flow through the nozzle orifice. An increase or decrease in pressure affects the size of spray droplets; increasing the pressure produces a finer spray with potential drift hazard.

Pumps are operated by gas or air pressure and liquid pressure. In gas pressure, the entire system is under pressure. The spray tank is pressurized to force the liquid from the tank through the nozzles. In this system, liquid air, liquid CO_2 or gases such as ammonia can be released through regulating valves to provide gas pressure for the sprayer. In liquid pressure systems, the spray tank is not pressurized. Part of the liquid is by-passed back to the tank for agitation and for pressure control.

The principal types of pumps used in spraying machines are piston pump, centrifugal or turbine pump, gear pump, impeller pump and diaphragm pump.

Filters and Strainers

Generally filters or strainers are built as a part of the nozzle, placed on the inlet of the spray tank or installed as a line filter. In filling the tank, a coarse strainer also should be used to filter the spray fluid. This is essential if the water contains trash or dirt. Wettable powders rarely pass through screens finer than 50 mesh. Screens for emulsions should be about 100 mesh.

Pressure Regulators

A pressure regulator or relief valve is used to maintain a relatively constant desired pressure. The valves generally consist of a ball pressurized by a spring, which is screw-adjusted to vary the pressure. Variations include a double-spring type and a combination of spring and diaphragm. Excess liquid is released back into the spray tank through a bypass port. A pressure gauge is usually placed on the outlet side of the regulator to indicate nozzle pressure.

Spray Boom

A boom is a horizontal pipe to which the nozzles are ûtted at the required spacing. Its length varies from 1 to 10 m, depending on the sprayer used. For hand-operated sprayers, a boom is normally fitted with 2 or 3 nozzles. More area is covered with a boom in one trip because of its wider swath.

Hoses

Hoses should be of a material that will withstand the chemicals used, particularly the petroleum solvent. Synthetic rubber and plastic are the materials most commonly used in spray hoses.

Spray Lance

Made of brass with a minimum diameter of 6 mm, the normal length of the lance of a hand-operated knapsack sprayer is 1 m. One end of the spray lance is fitted with a hose and the other with a nozzle. The nozzle end is normally bent to form a goose neck. A spray boom containing more than one nozzle can be fitted to the lance either horizontally or vertically. At the joint of the hose and lance, a trigger mechanism is fitted to shut off flow of the liquid. When necessary, a spray shield can be fitted at the delivery end of the lance (Fig.2.5.5).

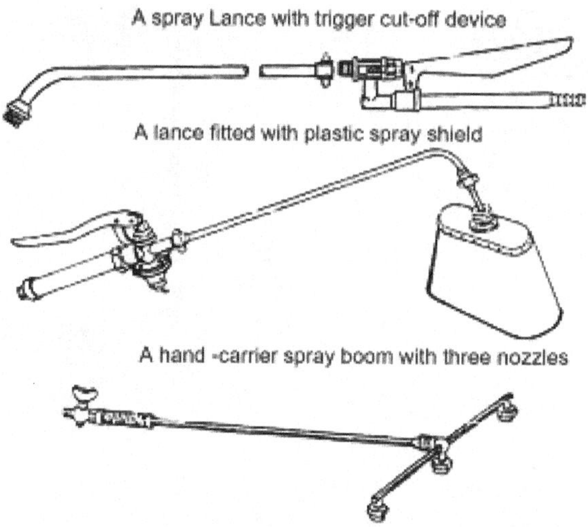

Fig. 2.5.5: Spray lance with different fittings

Types of Herbicide Application Equipment

Sprayers

High volume sprayers

1. Hand-operated Sprayers

Knapsack/ Bakpak *Sprayers*

a) Hydraulic Knapsack

Hydraulic knapsack sprayer with a capacity of 5-15 litres is the most useful and widely used sprayer in India for herbicide application. It can be carried on the back with help of straps. It is provided with a double action lever operated pump, mounted either inside or outside the container. The operator operates the lever with his left hand and pumps the spray liquid to the lance held by his right hand. The spray liquid is drawn from the tank by a suction stroke through a non-return valve and then forced through a delivery valve into the pressure chamber by a return stroke. The air in the pressure chamber is then compressed due to the liquid charged by the pump. When the cutoff valve is operated, the air, which got pressurized in the pressure chamber, forces the liquid out through the discharge hose and nozzle. Pumping is to be done almost continuously to maintain uniform pressure.

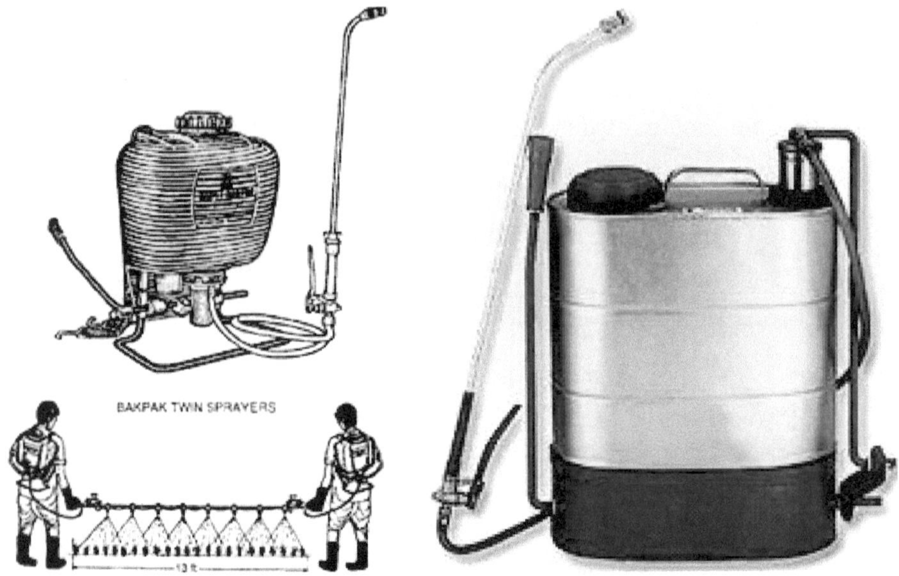

Fig. 2.5.6: Left=A hydraulic, knapsack sprayer with a plastic tank and spray lance. Right=A hydraulic, knapsack sprayer with a Brass tank and spray lance. Bottom=A manual device for obtaining multi-nozzle, spray boom action (diagrammatic). The arrangement requires harmonious movement of the two workers.

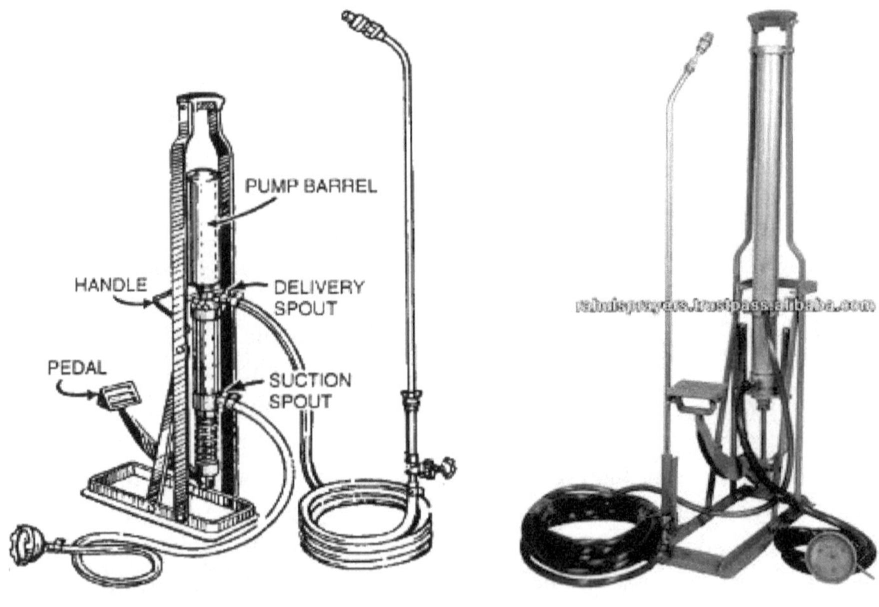

Fig. 2.5.7: A foot sprayer, or pedal pump sprayer, with one delivery hose. When required it has also provision to use two delivery hoses to hasten the work.

b) Pneumatic Knapsack Sprayer

The pneumatic or compressed system knapsack sprayer does not require pumping during spraying. The tank is pressurized after filling it with liquid with a built-in pump, a separate charge pump or CO_2 cylinders. The main disadvantage of this sprayer is that uniform delivery pressure cannot be maintained, resulting in an uneven delivery. It is more useful to spray on hilly terrain and irrigated fields where walking is difficult.

Fig. 2.5.8: Pneumatic knapsack sprayer

c) Logarithmic Sprayer

A logarithmic sprayer begins spraying at a high rate and decreases the concentration logarithmically. The sprayer has two tanks which are connected through tubing and valves. One tank is filled with herbicide and the other with water. The water flows from the diluter tank into the herbicide tank and form this tank through the nozzles. Initially, the liquid passes through the nozzles at maximum concentration. As the liquid in the herbicide tank gets diluted logarithmically with water from the diluter tank, lower and lower concentrations are sprayed.

The logarithmic sprayer is very useful for experimental work in herbicide evaluation to establish crop and weed tolerance to a chemical at varying rates.

2. Motorised (Power) sprayer

Power sprayers are fitted with a 35cc or 60-70cc engines. A direct drive connects the engine to a centrifugal fan, which is usually mounted vertically to a strong L shaped frame (Fig.2.5.9). The fan produces a high velocity air stream, which is diverted through a 90 degree elbow to a flexible discharge hose. At the end of which is mounted a nozzle and the spray fluid is discharged directly from the tank. Here the fluid gets atomized and spraying is effected. Engine speed is often 6000 rpm. The spray tank is mounted above the engine/fan unit and normally holds about 10 litres.

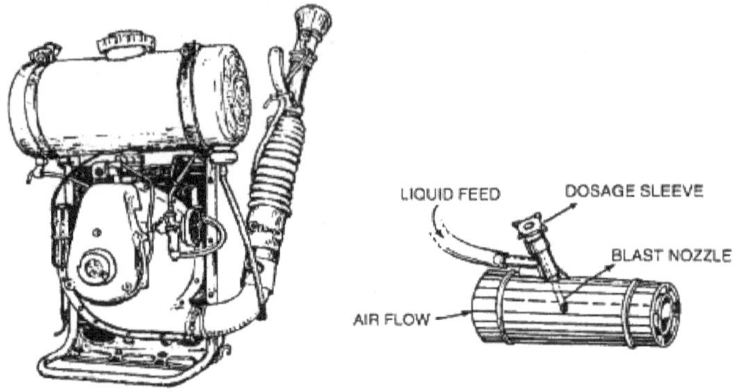

Fig. 2.5.9: Left=A motorized knapsack sprayer. Right=Spraying jet system of a motorized knapsack sprayer.

3. Micron Herbi sprayer

Controlled Droplet Application

The technique of controlled droplet application (CDA) is used to make ultralow volume (ULV) spraying of herbicides possible. CDA is defined as the production and application of droplets that are the appropriate size both for the target and the method of delivery. In the CDA technique of spraying of ultralow volume, the production of droplets is controlled within very close limits and the size is tailored to ensure that the herbicide in the spray mixture has the maximum effect on the target at which it is directed. Because of this, the amount of spray volume required is dramatically reduced to the extent that it needs to be applied only at ultralow volumes.

The controlled size droplets are produced by rotary-atomizer sprayers which are now widely known as ULV sprayers or controlled droplet applications. The CDA technique is used to create a fine mist which drifts onto all exposed areas of plants, so that they become well covered with myriads of droplets of the herbicide. In the case of herbicides, CDA is done using a ULV sprayer named 'Herbi' or 'Micron Herbi' (Fig.2.5.10). The atomizer is driven by a small electric motor which is attached to a tube carrying the batteries that provide power for the motor. The spray volume for Herbi is between 7 and 15 L ha^{-1}.

Herbi covers a swath of 1.2 m wide with a walking speed of 3 to 5 km ha^{-1} or one to two paces per second. The discharge rate is approximately 1 m s^{-1}.

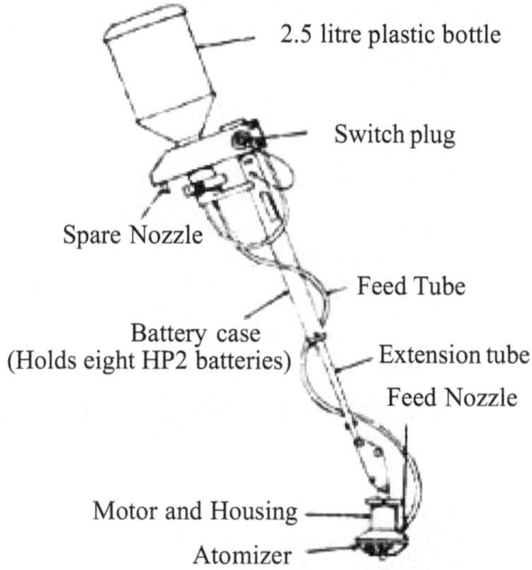

Fig. 2.5.10: The Micron Herbi sprayer

The CDA method is particularly useful in the case of foliage-applied systemic herbicides such as glyphostate, 2, 4-D, MCPA, mecoprop, dicamba, dichlorprop, asulam, etc. However, contact herbicides such as ioxynil, bromoxynil, paraquat, etc. are less effective when applied by CDA.

4. Power-driven Sprayers

The most common type of power equipment for spraying herbicides is that which uses a hydraulic pump of plunger, rotary or centrifugal type. The fluid is forced through nozzles which break it up into droplets and direct the droplets at the surface to be covered.

Power-driven sprayers are usually mounted on a tractor on the three-point linkage. Tractor sprayers are equipped with a spray boom although sometimes clusters of nozzles are used (Fig.2.5.11) The spray from a boom is less affected by wind than that from the nozzles cluster, but unevenness of the ground can be more serious as the movement is magnified along the length of the boom. This can be minimized by a good design and by using full cone jets rather than hollow cone or fan jets.

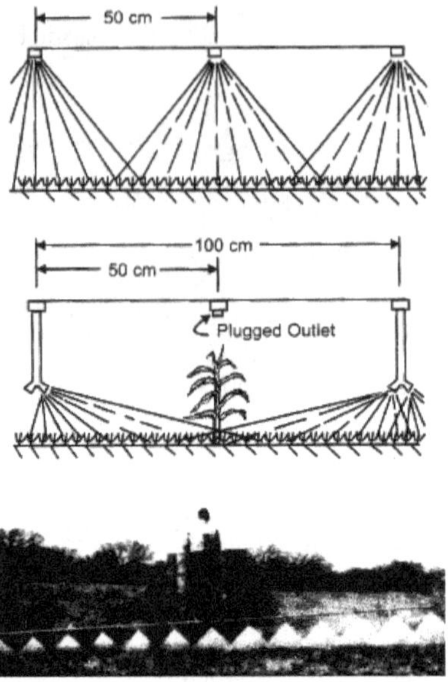

Fig. 2.5.11: A tractor- mounted spray boom fitted in two fashions. Top=spray boom fitted too low, causing overlap sprays. This should be avoided. Middle =Spray boom well adjusted to avoid spray falling on the crop foilage but wet the weeds completely without any gaps or overlpas (diagrammatic). Bottom-A tractor= mounted sprayer in action with its boom tilted, by fault. The result is a gappy spray pattern towads the left and undue overlap sprays towards the right. The spray pattern from the middle three nozzles alone is appropriate (Butter and Bowers, 1961).

Fig. 2.5.12: Aerial application of herbicide

Fig. 2.5.13: Modern applicant to apply herbicides

5. Direct Contact Applicators

Direct Contact Application (Dca)

In the DCA method, the herbicide is placed, wiped, rubbed or smeared onto the plant surface. DCA offers the following benefits : greater speed and ease of operation than hand weeding, freedom from concern about walking speeds as contrasted to knapsack spraying operations, ability to treat areas inaccessible to other types of cultivation or herbicide application equipment, possibility of effectively controlling weeds that are difficult or impossible to hand weed due to hard-to-reach propagating organs and/or a thorny, bristly or barded nature, reduced volume of herbicide required due to treating only the target plants, and potential for local fabrication of most parts plus assembly. DCA can be achieved by re circulating sprayer, rope wick applicator, roller application and herbicide glove.

Recirculating Sprayer

The recirculating sprayer has the advantage of the spray being directed horizontally only to weeds growing above the crop so that minimum amounts of herbicide contact the crop plants. The herbicide spray that is not deposited on weeds is collected in the re circulating sprayer and re-applied.

Rope-Wick Applicator

The rope-wick applicator, invented by].E. Dale of the United States Department of Agriculture (USDA), Stoneville, Mississippi, USA, uses capillary action and gravitational flow to move the chemical out of a reservoir onto nylon rope wicks where the chemical is rubbed from the soaked rope onto the tall weeds. This method of application not only eliminates any possibility of trash getting into the herbicide reservoir but also greatly reduces the possibility of injuring crop plants by herbicide drift, spillage and splattering. The rope wick acts much like the wicking system used in kerosene lamps to deliver fuel to the flame. As the applicator moves over the field, the herbicide is applied only on weeds contacted by the soaked ropes. As a result, very little chemical is used for treating weeds on one hectare. Rope-wick application has been found particularly useful for systematic foliage herbicides such as glyphosate, 2, 4-D, etc.

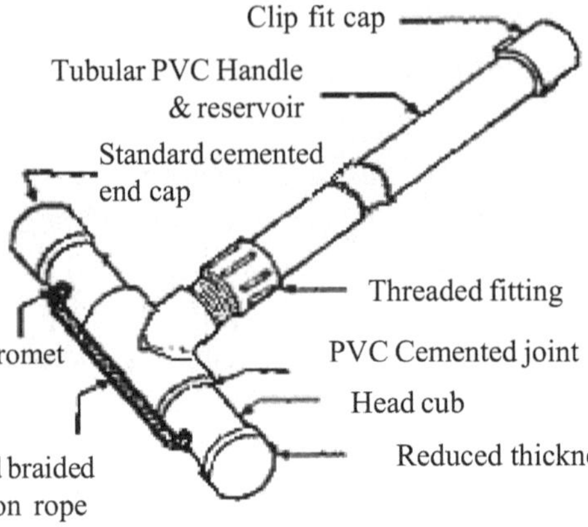

Fig. 2.5.14: A prototype Rope-Wick Applicator

Roller Applicator

Like the re circulating sprayer, the roller applicator applies herbicide only to weeds growing taller than the crop. The roller applicator wets weeds better than the rope-wick applicator but less effectively than the re circulating sprayer. Wind does not affect application with a roller applicator as it does with the re circulating sprayer. However, the roller applicator has a tendency to drip herbicide onto the crop when operated at an angle in hilly terrain and to pick up prickly weed seeds.

Herbicide Glove

The herbicide glove is designed to allow safe application of specifically effective herbicides to tough-to-control weeds. It facilitates exclusive application of the herbicide to the weed. The herbicide glove consists of a plastic belt from which the container is suspended, an active glove for one hand, with flow tube, control valve, pressure bulb and sponge or foam pad, and a neutral glove for other hand (Fig.2.5.15). All the parts are made of tough plastic. Below the protective palm and foam pad rests a rubber bulb which is deflated with each grip or squeeze of the glove. While walking along the crop row, the stem just below the panicle of each weed is squeezed to deposit the spray on the plant. Each squeeze of the hand releases about 1 ml of the herbicide in the palm pad.

The herbicide glove method is slow but can help control weeds selectively and effectively. This is more useful for control of such weeds as wild oats and many tall weeds.

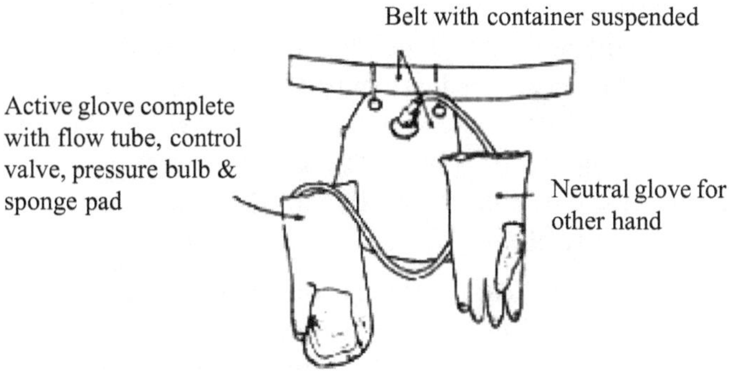

Fig. 2.5.15: Herbicide glove

Electrodynamic Spraying

In electro dynamic spraying, the spray liquid is atomized by an electric field. As a result, the droplet is charged and the spray liquid propelled away [from the nozzle onto the target.

Fig. 2.5.16: Electrodyn, battery–operated sprayer

It is used for treating small plots. The liquid is fed by gravity to a nozzle to which a high positive charge, which is capable of spreading out the herbicide droplets on the target vegetation, is applied.

Granular Application

Granular formulation is increasingly being used particularly in the case of PRE herbicides as it offers the following advantages against spraying: (a) water is not needed to apply the herbicide, (b) equipment cost is reduced and (c) crop injury due to spray drift is non-existent.

Granular Applicator

With the rapid increase in usage of granular formulations of PRE herbicides, there has been a corresponding increase in the popularity of granular applicators.

They broad cast the granules over the entire crop row or in hands. A granular applicator consists of the hopper in which the granules are placed, a rotating disc device to get the broadcast application pattern, a long flexible discharge tube with a nozzle at the distal end and a finger-controlled mechanism for regulating the flow of granules. By rotating the handle, the granules are released into the tube through the exit hole of the hopper. The applicators are made of plastic and have a capacity of 1 to 10 kg. They are carried on the back.

The bigger type granule applicators are tractor-mounted. They consist of a band distributor or spatter plate that distributes granules laterally and uniformly over the desired band width and an easily calibrated device that meters the granules.

Fig. 2.5.17: Granular applicator

Sand Mix Application

Granules of herbicides can be mixed in sand and applied, broadcast, on the soil. This is a widely used method for PRE and E-POST (early-POST) application of granular formulation in India and other rice-growing countries in irrigated transplanted as well as direct seeded rice.

Dry sand of fine- to medium-coarse texture is preferred for mixing the herbicide granules. Sand granules should be nearly of the size of herbicide granules to avoid the segregation problem during mixing and application. Thorough mixing of sand, 50-75 kg ha^{-1}, and the required quantity of herbicide granules is done on clean cemented ground. The herbicide sand mixture is then broadcast uniformly over the field by experienced workers. Uniformity in application enhances weed control efficacy. Herbicides such as butachlor, 2, 4-D ethyl ester, anilophos, thiobencarb, propanil, and other rice herbicides can be used as sand mix. Numerous studies conducted in farmers fields showed that there was no reduction in the efficacy of weed control by using herbicide granule-sand mixture.

Sand mixing reduces the cost of application to the minimum. It also helps in saving time and energy required for herbicide application.

Herbicide Incorporation

Soil-applied herbicides which have medium to high volatility require incorporation into the soil. Herbicides such as trifluralin, benefin, EPTC, vernolate, dinitramine, pebulate, bensulide, profluralin, butralin, etc. require thorough mixing with the soil for effective weed control. Incorporation not only prevents their loss by volatilization or photodecomposition by sunlight, but also reduces the soil moisture requirement and places the chemical in close contact with seed or roots for best control.

Mixing of herbicides in the soil is done by incorporating implements, e.g. spring tooth harrow, bed conditioner, rotary hoe, rolling cultivator, pulverizing tandem disc harrow, etc. The most popular power-drawn incorporating tool in the USA is the pulverizing tandem disc harrow. In this case, the herbicide is applied from a spray boom mounted on the tractor. A disc which cuts 10 to 15 cm deep provides effective incorporation on most soil types. The spring-tooth harrow is also a useful tool as it cuts as deep as 15 cm except in heavy clay soils.

Herbicide Application Through Irrigation

Application of herbicides in irrigation water (chemigation) is a relatively new phenomenon. Herbicides may be applied through surface (furrow or flooding) irrigation, drip or trickle system and sprinkler system.

Surface Irrigation System

The surface irrigation system, also termed the gravity-flow system, is used in case of bed-planted furrow- or flood-irrigated crops. Herbicides can be applied immediately after planting, as PRE treatment, during the first irrigation. The condition of the furrow is important because it influences rate of water ûow, amount of water applied and uniformity of distribution. Hence, a clod-free furrow is very helpful in controlling water flow.

In coarse-textured soils, water application of 4-5 cm is required for complete wetting of the soil, while in fine-textured soils, 8-10 cm water is needed. Very sandy soils are not well suited for this type of herbicide application. In this gravity flow system, a minimum slope of 0.2% is needed to achieve uniform water application.

Sprinkler System

The sprinkler system is the more widely used irrigation system for herbicide application. The calibration procedure is similar to the one described for the surface irrigation system.

The sprinkler system is more useful when herbicides are applied PRE to weeds, but not necessarily to crops. Herbicides that can be used include alachlor, atrazine, benefin, bromoxynil, butylate, EPTC, metolachlor, metribuzin, pendimethalin, vernolate, certain substituted ureas and sulphonyl ureas, etc.

Drip or Trickle System

In this system, water is applied to the soil through orifices or emitters located at selected points along the water delivery lines. As herbicides applied through the drip or trickle irrigation systems do not move in the soil as far as the irrigation water, weeds grow in the non-treated moist area near the weed-free area around the emitter. The size of the weed-free area is affected by length of the irrigation period, rate of water application and concentration of the herbicide.

In the drip irrigation system, the amount of herbicide in solution is expressed in ppm. For deep-rooted perennial crops, herbicide concentrations of 20-100 ppm are applied over 2-4 h.

Equipment for Herbicide-Irrigation

The basic equipment for delivering the herbicide into an irrigation system includes a chemical supply tank, an injection system, and the appropriate safety and anti-siphon devices that prevent potential contamination of the water source.

Application of Herbicides in Fertilizers

Fertilizers offer an alternative method of application of herbicides, particularly the soil applied ones. Dry fertilizers can be used as a carrier for PRE herbicides such as dinitro anilines, triazines, ureas, thio carbamates, sulfonyl ureas, etc. Dry fertilizers became popular in the late 1960s in the USA when thio carbamates were impregnated on the fertilizers used. Since then, several other herbicides have been used for impregnating bulk blended fertilizers. Similarly, liquid fertilizer as a carrier for herbicides was used in the USA in the late 1950s. The combination of nitrogen solution and triazine herbicide was the first mixture to gain wide acceptance.

Herbicide Coating of Crop Seed

Crop seeds are treated with various chemicals such as fungicides, insecticides, nematicides and herbicide safeners to protect seeds and emerging seedlings from soil borne insects, diseases, nematodes and herbicides. Besides, seeds are also pelleted or coated with minerals, carbon protectants and inoculum of nitrogen-fixing bacteria. In view of these examples, coating of crop seeds may seem to be a convenient method of herbicide application to control weeds in the vicinity of the germinating seeds and emerging seedlings.

Time of Application of Herbicides

Based on the time of application of herbicides, the methods are classified in to pre-planting, pre-emergence and post-emergence.

Pre-planting: Application of herbicides before the crop is planted or sown. Soil application as well as foliar application is done here. For example, Fluchloralin can be applied to soil and incorporated before sowing rainfed groundnut while Glyphosate can be applied on the foliage of perennial weeds like *Cyperus rotundus* before planting of any crop.

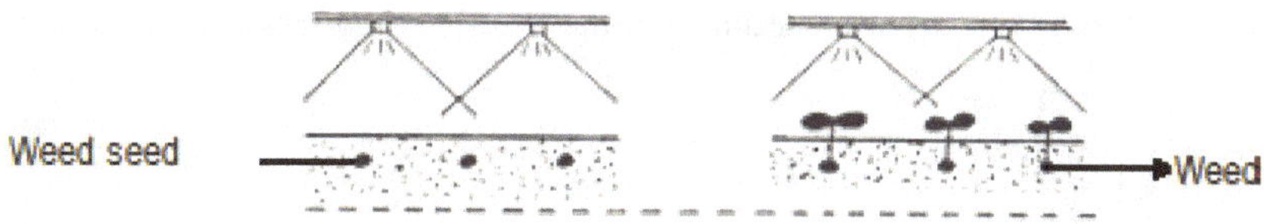

Pre-emergence

Application of herbicides before a crop or weed has emerged. In case of annual crops application is done after the sowing of the crop but before the emergence of weeds and this is referred as pre-emergence to the crop while in the case perennial crops it can be said as pre-emergence to weeds. For example soil application by spraying of Atrazine on 3rd DAT to sugarcane can be termed as pre-emergence to cane crop while soil application by spraying the same immediately after a rain to control a new flush of weeds in a inter-cultivated orchard can be specified as pre-emergence to weed.

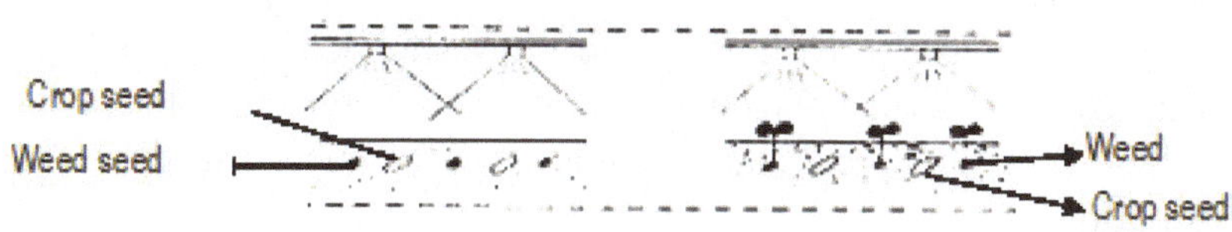

Post-emergence

Herbicide application after the emergence of crop or weed is referred as post-emergence application. When the weeds grow before the crop plants have emerged through the soil and are killed with a herbicide then it is called as early post-emergence. For example spraying 2,4-D Na salt to control parasitic weed striga in sugarcane is called as post-emergence while spraying of Paraquat to control emerged weeds after 10-15 days after planting potato can be called as early post-emergence.

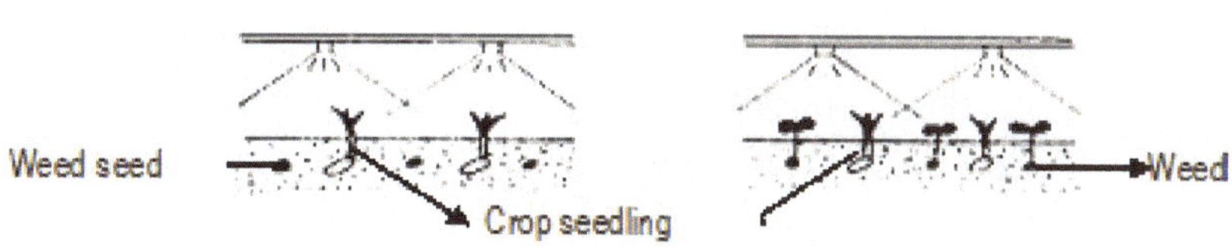

Fig.2.5.18: Post-emergent application of herbicide using early application equipment

Spray Precautions

Spray Drift

Spray drift usually occurs when the droplets are blown by the wind during spraying. This can be minimized by (a) spraying when no wind is blowing, (b) using suitable sprayer, (c) lowering the spray height from the ground, (cl) using proper nozzles and (e) reducing the delivery pressure.

At high delivery pressure, the droplet size becomes smaller and the smaller droplets are readily carried away by wind. Suitable pressure-regulating or pressure-control valves are useful in maintaining a constant pressure at the outlet regardless of the inlet pressure. Pressure valves need frequent checking. Low-drift jets producing very few small droplets help in reducing the drift.

In order to prevent drift completely, spray shields are very useful. The shield must match the nozzle to ensure that the entire shielded area is sprayed. The shield is, however, disadvantageous because the operator cannot see the spray jet to direct it properly, and also dripping from the edges causes wastage of spray solution.

Spray Mixing

The effect of herbicide at a particular rate of application depends largely upon mixing the correct quantities of herbicide and water. Therefore, one must determine the quantity of spray volume required per unit area. This can be done by spraying a measured quantity of water on a known area and calibrating the actual quantity required for 100 metres, 1000 metres or one hectare. This is done as shown under.

Volume applied per hectare

$$= \frac{\text{Water used (litres) x10, 000}}{\underset{\text{(metres)}}{\text{Distance covered}} \text{ x } \underset{\text{(metres)}}{\text{width of spray swath}}}$$

For example, if 80 meters were covered while s praying solution of 3.6 litres with a swath width of 1 metre, the volume applied per hectare would be:

$$\frac{3.6x10,000}{80\times 1}=\frac{36,000}{80}=450 \text{ litres}$$

Once the spray volume rate is determined, one must calculate the exact amount of commercial formulation of a herbicide. The formulations contain less than 100% of the active ingredient (ai) and as the rates are usually quoted in terms of active ingredient the formulated product required is calculated thus:

$$\frac{\text{Rate of ai/ha to be applied x100}}{\text{Active ingredient (\%)}}$$

For example, if one needs to apply diuron, at 2 kg ai/ha and the commercial formulation contains 80% active ingredient, the total product required per hectare would be:

$$\frac{2x100}{80}=2.5 \text{ kg}$$

For some compounds, the activity is expressed on the basis of per cent concentration of the parent acid in the formulated product. In the case of Gramoxone, a commercial formulation of paraquat dichloride salt, the active ingredient in the form of paraquat dichloride is 24% and in the form of parquat ion 20%.

The wettable powders and emulsifiable concentrates should be initially mixed with an equal amount of water to ensure complete wetting and then brought to the required concentration by adding the remaining quantity of water. Perfectly clean water should be used for herbicide spraying.

Assessment of Herbicide Efficacy

Herbicides may be assessed both for its effectiveness and phytotoxicity as described below:

Effectiveness

For assessing the effectiveness of an herbicide, observation on weed population and dry matter accumulation should be recorded at speciûed intervals (weekly, fortnightly). This should be done at least from two quadrates in each sampling plot.

Indices for the Efficiency of Herbicides

Some of the indices worked out to ûnd out the efficacy of a particular method of weed control are:

Weed Control Efficiency (WCE)

$$WCI = \frac{Wc\text{-}Wt}{Wc} \times 100$$

Where Wc is the average weed count or weight per unit area in the unweeded check plot, and Wt is similar information obtained from the treated plots.

$$\text{Weed Index (WI)} = \frac{Y_{WF}\text{-}Yt}{Y_{WF}} \times 100$$

Where Y_{WF} is the average crop yield in the weed free plot and Y_t is the average crop yield in the herbicide treatment under study.

In experiments dealing with the determination of the efficacy of some inter-crop on the suppression of weeds vis-a-vis sole cropping, the WCE in the first formula is replaced by. Weed Smothering Efficiency (WSE) which is given as:

$$WSE = \frac{Mdw\text{-}ldw}{Mdw} \times 100$$

Where Mdw and ldw are the average dry weight obtained from the sole main crop plots and the intercropped plots, respectively.

Weed Control cum Smothering Efficiency (%)

$$WCSE = \frac{\text{Weed dry weight in control cum sole crop (kg/ha)} - \text{Weed dry weight in weed management cum smothering intercrop (kg/ha)}}{\text{Weed dry weight in control cum sole crop (kg/ha)}}$$

Phytotoxicity

Phytotoxicity of a particular herbicide could be assessed by taking observations on seed germination (in the case of pre-emergence herbicides), plant height, number of tillers (in tillering crops), dry matter production and ûnal grain and straw yield. Phytotoxicity of herbicides may also be assessed by visual observations, particularly for post-emergence herbicides. These observations may be taken by more than 3 persons and the average is arrived for proper assessment. Scoring may be done from 1-10 for making comparative assessment (Table 2.5.1).

Table 2.5.1: Rating Scale

Injury (%)	Rating
1-10	1
11-20	2
21-30	3
31-40	4
41-50	5
51-60	6
61-70	7
71-80	8
81-90	9
91-100	10

In addition to assessing the damage to the foliage, it is also important to note down the time required for recovery.

2.6

Mode of Action, Translocation and Absorption of Herbicides

Mode of Action

The term mode of action refers to the sequence of events from absorption into plants to plant death. The mode of action of the herbicide influences how the herbicide is applied. For example, contact herbicides that disrupt cell membranes, such as acifluorfen (Blazer) or paraquat (Gramoxone Extra), need to be applied post emergence to leaf tissue in order to be effective. Seedling growth inhibitors, such as trifluralin (Treflan) and alachlor (Lasso), need to be applied to the soil to effectively control newly germinated seedlings.

To be effective, herbicides must 1) adequately contact plants; 2) be absorbed by plants; 3) move within the plants to the site of action, without being deactivated; and 4) reach toxic levels at the site of action. The application method used, whether pre-plant incorporated, pre-emergence, or postemergence, determines whether the herbicide will contact germinating seedlings, roots, shoots, or leaves of plants.

Mode of Action of Herbicides

Herbicides perform a vital role in the management of weeds. As the name indicates, herbicides are chemicals that kill or control vegetation. Although the ultimate effect of most herbicides is the same (usually death of weed), the way they control weeds is vastly different. Physiologists use the term mode of action to describe the way the herbicides affect weeds. It includes the entire sequence of events that occur from the time the weed absorbs the herbicide to the final plant response (usually death). The term mode of action is the broad term under which all aspects of herbicidal action including the mechanism of action is included, while the mechanism of action refers to only the biochemical and biophysical responses of plants that appeared to be associated with herbicidal action.

Thus, mode of action includes absorption, translocation to an active site, inhibition of a specific biochemical reaction, degradation or breakdown of the herbicide in the plant and soil and the effect of the herbicide on plant growth and physiology.

Although two herbicides may differ chemically, they may still possess the same mode of action example trifluralin (a dinitroaniles herbicide) and propanamide (an amide herbicide) are inhibitors of microtubule/spindle apparatus. Each herbicide family (class or group) has a primary site of action which may be different in its action from others example sulfonylureas herbicides are ALS or AHAS inhibitors while glyphosate and sulfosate are ESPS inhibitors. Some may have more than one site of action, but the mostinhibitory of these will be affected first. The other site(s) may be considered secondary. Fluometuron, a urea group herbicides act by inhibiting photosynthesis at photosystem II and carotenoid biosynthesis.

The primary mode of action of major groups of herbicides which can sufficiently explain their phytotoxicity through their effect of biological processes, are reviewed as below.

1. **The Growth Regulator Herbicides:** (2, 4-D, MCPP, dicamba, and triclopyr). These are mostly foliar applied herbicides which are systemic and translocate in both the xylem and phloem of the plant. They mimic natural plant auxins, causing abnormal growth and disruption of the conductive tissues of the plant. The injury from this family of herbicides consists of twisted, malformed leaves and stems.
2. **The inhibitors of amino acid synthesis** (glyphosate, halosulfuron, imazethapyr, and sulfometuron). Both foliar and soil applied herbicides are in this family. Glyphosate translocates in the phloem with photosynthate produced in the leaves. Others in this family move readily after root or foliar absorption. These herbicides inhibit certain enzymes critical to the production of amino acids. Amino acids are the building blocks of proteins. Once protein production stops, growth stops. Symptoms are stunting and symptoms associated with lack of critical proteins.
3. **Cell membrane disrupters** with soil activity (oxyfluorfen, lactofen, and acifluorfen). Soil and foliar applied with limited movement in soil. These herbicides enter the plant through leaves, stems, and roots, but are limited in their movement once they enter the plant. Membrane damage is due to lipid peroxidation. Symptoms are necrosis of leaves and stem.
4. **Lipid biosynthesis inhibitors** (diclofop, fluazifop, sethoxydim, and clethodim). Foliar applied Diclofop has both soil and foliar activity. Herbicides in this family move in both the xylem and phloem of the plant and inhibit enzymes critical in the production of lipids. Lipids are necessary to form plant membranes which are essential to growth and metabolic processes. Symptoms include stunting and death of tissue within the growing points of plants.

5. **Pigment inhibitors** (norflurazon, fluridone, and amitrol). Soil applied and moves in the xylem except amitrol, which moves in both phloem and xylem. These herbicides inhibit carotinoid biosyntehsis, leaving chlorophyll unprotected from photooxidation. This results in foliage which lacks color. Symptoms include albino or bleached appearance of foliage.
6. **Growth inhibitors of shoots** (thiocarbamate herbicides including: EPTC, cycloate, pebulate, and molinate). Soil applied and somewhat volatile, requiring incorporation. Enter the plant through the roots and translocated through the xylem with the transpiration stream to the growing points in the shoot. Mode of action is unclear, but affects developing leaves in growing points of susceptible plants. Symptoms include stunting and distortion of seedling leaves.
7. **Herbicides which disrupt cell division** (trifluralin, DCPA, dithiopyr, oryzalin, pronamide, pendimethalin, and napropamide). All are soil applied with limited movement in the soil are absorbed through roots or emerging shoot tips. Once absorption takes place, movement is limited (site of action is near the site of absorption). These herbicides inhibit cell division or mitosis, except pronamide and napropamide which stop cell division before mitosis. Symptoms include stunting and swollen root tips.
8. **Cell membrane disrupters** no soil activity (paraquat, diquat, glufosinate, acids, oils, soaps). These herbicides are foliar applied with no soil activity. They enter the plant through the leaves and stems and do not move significantly within the plant once absorbed. These herbicides either act directly on cell membranes (acids, soaps. oils) or react with a plant process to form destructive compounds which result in membrane damage. Symptoms include rapid necrosis of the leaves and stem.
9. **Inhibitors of photosynthesis** (atrazine, simazine, metribuzin, cyanazine, prometryn, diuron, linuron, tebuthiuron, and bromacil). These are soil applied herbicides. However, all except simazine also have foliar activity. They move readily in the plant in the xylem with the transpiration stream where they concentrate in the leaves at the site of photosynthesis. Once there they block the electron transport system of photosynthesis, causing a buildup of destructive high energy products which destroy chlorophyll and ultimately the leaf tissues. Symptoms include chlorotic (yellowed) leaves which become necrotic.

Butachlor

Butachlor is the first herbicide in pre-emergence chemical group to be commercialized,a Chloroacetamide herbicide.

Suitable crops: rice

Mode of action: Very long fatty acid inhibitors.

Mechanism : Long-chain fatty acid:CoA ligase is inhibited by the carbamothioate herbicides EPTC and triallate. However, details about other elongases inhibited by these herbicides are unknown.

Pretilachlor

Technical Name:Pretilachlor 50 % EC, Brand Name:PRINCE 50EC

Mode of action : Pretilachlor belongs to the chloroactamide class of herbicides, which inhibits growth and reduces cell division. It offers reliable pre-emergence to early post-emergence control of annual grasses, some sedges and broadleaf weeds for high yield in transplanted and dry-sown rice cropping systems. It is especially designed for early season weed control in wet-sown rice and rice nursery beds.

Apart from classifyingherbicides according to similarities of structures, these can also be divided into differentgroups depending upon their similarities in their mode of action as given in Table 2.6.1.

Table 2.6.1: Herbicides mode of action group according to the riskfor early development of resistance.

Group A	**Inhibition of fat (lipid) synthesis - ACC'ase inhibitors** Aryloxyphenoxy propionates (Fops), Cyclohexanedionesoximes (Dims)
Group B	**Inhibition of the enzyme acetolactate synthase -ALS inhibitors** Sulphonylureas, Imidazolinones, Sulphonamides
Group C	**Inhibition of photosynthesis at photosystem II**Triazines, Triazinones, Ureas, Nitriles, Benzothiadiazoles, Acetamides,Uracils, Pyridazinones, Phenyl-pyridazines
Group D	Inhibitors of tubulin formationDinitroanilines, Benzoic acids
Group E	**Inhibitors of mitosis**Thiocarbamates, Carbamates, Organophosphorus
Group F	**Inhibitors of carotenoid biosynthesis** Nicotinanilides, Triazoles, Pyridazinones, Isoxazolidinones , Pyrazoles
Group G	Inhibitors of protoporphyrinogen oxidaseDiphenyl ethers, Oxadiazoles
Group H	**Inhibitors of protein synthesis**Thiocarbamates
Group I	**Disrupters of plant cell growth**Phenoxys, Benzoic acids, Pyridines
Group J	**Inhibitors of fat synthesis**Alkanoic acids
Group K	**Herbicides with multiple sites of action**Amides, carbamates, Amino propionates, Benzofurans, Phthalamates, Nitriles
Group L	**Inhibitors of photosynthesis at photosystem I**Bipyridyls
Group M	**Inhibitors of EPSP synthase**Glycines
Group N	**Inhibiters of glutamic synthates**Glycines

Groups listed by resistance risk: A to B-high risk, C to H-Moderate risk, I to N- Low Risk. *Source*-Grains Research and Development Corporation (1999)

Inhibition of Photosynthesis

Photosynthesis involves the conversion of light energy into chemical energy, the lightreactions, and the conversion of carbon dioxide into sugars, the dark reactions. The light reaction captures light energy and converts this into chemical energy through the electron transport chain. The products of light reaction are chemical energy in the form of ATP, reducing power in the form of NADPH and oxygen as a by-product. The light reactions are divided into two cycles: Photosystem l (PSI) or cyclic phosphorylation and Photosystem II or non-cyclic phosphorylation. Both involve the capture of light energy by chlorophyll, a photo receptor and the acceptance of electrons from the splitting of water (Fig.2.6.1). The capture of these electrons increases the energy level of the chlorophyll to the so-called single state and then returns to the ground state as the electrons flow through an electron transport chain to produce ATP and NADPH. If the electron transport chain is interrupted and light continues to fall on the chloroplast, energy level of the chlorophyll is raised from singlet state to triplet state. Triplet chlorophyll can interact in a damaging way with membrane lipids but, more importantly, it can excite oxygen, there in abundance because of photosynthesis, to a single state. This singlet oxygen is very reactive, and it interacts with cellular lipids, proteins, nucleic acids and many other plant cell components there by inducing cellular disorganisation and plant death.

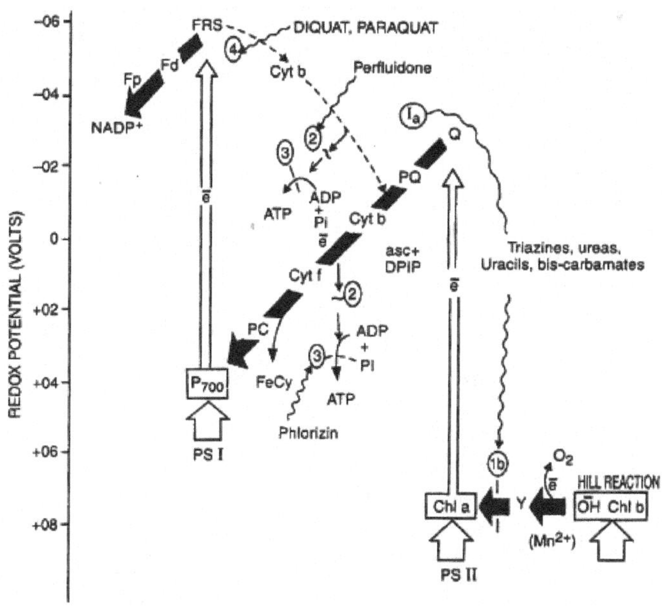

Fig. 2.6.1: Schematic presentation of the electron transport and phosphorylation reactions considered to occur in chloroplast lamella. The numbers 1a, 1b, 2,3 and 4 indicates the postulated sites of action of specific herbicides.

A number of herbicides interferes with the flow of electrons from singlet chlorophyllthrough the electron transport chain. The main herbicide groups are triazines, ureas,acylamides, hydroxy benzonitrile, phenyl carbamates, uracils and triazinones.

Inhibitor of PS II

The triazineblock electron transport in PS II by displacing QB from its binding site on theD_1 protein (Fig.2.6.2) (Fuerst and Norman, 1991). This action prevents the reduction of Q_B toPQH_2. In the absence of triazines, Q_B is bound to the D_1 protein by the hydrogen bonds : onebetween the Q_B carboxyl group and the hydroxyl group of serine (Ser)264 on the D1 proteinand the other between a second carboxyl group on Q_B and histidine (His)215 of the D_1protein (Fig.2.6.3). PS II electron transport inhibitors, such as atrazine block the Q_B bindingsite by binding to D_1 protein in the Q_B binding niche. Atrazine by means of hydrogenbinding, vander Wall's forces and hydrophobic reactions, binds to the D_1 protein at Ser264and phenylalanine(Phe) 265. Phe255 is also involved in hydrophobic interaction in thebinding of both Q_B and atrazine.

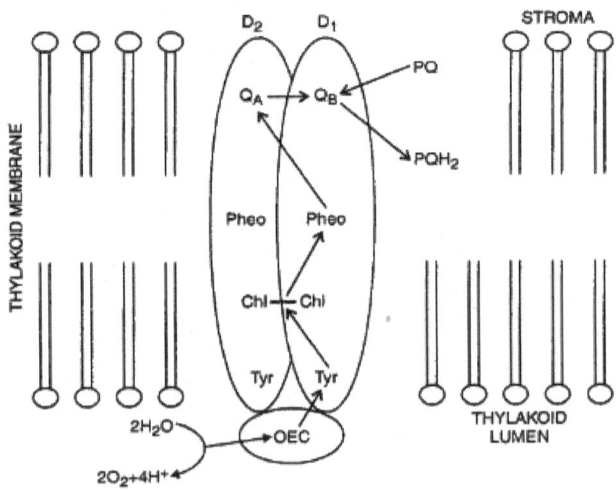

Fig. 2.6.2: PS II reaction centre complex. OEC, oxygen-evolving complex, Tyr, Tyrosine amino acidresidue 160; Chl-Chl, chlorophyll a dimer, also known as P680; Pheo, pheophytin; Q_A and Q_B,' bound plasloquinone; PQH_2, plastoquinone

Inhibition of PS I

The bipyridilium, paraquat and diquat divert the electron ûow of cyclicphotophosphorylation (PS I). Paraquat, a divalent cation, is a PS I electron acceptor that catalytically short-circuits light driven photosynthetic electron transport by accepting electronthat normally ûow to ferridoxin (Holt *et al.*,1993a). The paraquat radical reduces oxygen regenerating paraquat and forming the biologically reactive superoxide radical (Fig.2.6.3). While some superoxide is scavenged by superoxide dimutase and related protective enzymes, excessive superoxide results in production of highly toxic hydroxyl radicals. This toxic form of oxygen reacts with membrane fatty acids (lipid peroxidation) and other cellular components and leads to a rapid loss of membrane integrity, photooxidation, dessication andplant death. Sunlight, by enabling photosynthetic electron transport, greatly affects toxicity orparaquat.

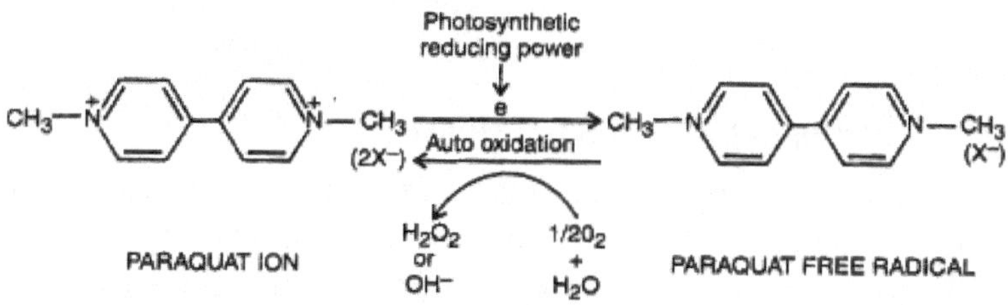

Fig. 2.6.3: Reversible, free radical formation from paraquat ion and its autooxidation yielding H_2O_2 or OH^-.

Inhibition of Chlorophyll Biosynthesis

Pyrazole herbicides such as benzofenap, pyrazolynate and pyrazoxyfen interfere withchlorophyll biosynthesis.Chlorophyll are the main photosynthetic pigments which are green in colour. Five maintypes of chlorophyll are found in the plants, out of which chlorophyll a (*Chl*a) is universal.Glycine and succinyl CoA react together in the presence of NAD, ATP and pyridoxylphosphate to form an unstable compound Q-amino â-keto-adipic acid. The latter loses carbondioxide and give rise to Q -amino levulinic acid, ALA. Two molecules of ALA condense in the presence of a dehydrase and produce a monopyrrole named prophobilinogen or PBG. Four molecules of PBG condense with a head to tail union resulting in the formation of a symmetrical ring of uroporphyrinogen III. The latter loses four molecules of carbon dioxide to give rise to coproporphyrinogen III. Coproporphyrinogen undergoes oxidativedecarbo-xylation and produces protoporphyrinogen IX. Protoporphyrinogen IX gets oxidizedby loss of six hydrogen atoms to protoporphyrin IX. It incorporates magnesium to form Mg-protoporphyrin IX.

Inhibition of Carotenoid Biosynthesis

In addition to the green chlorophyll pigments in the leaf in chloroplasts, there are otherpigments that can also capture light energy but that also protect the leaf from damagingradicals by quenching them. Carotenoids are examples of this type of pigment (Fig. 2.6.4).

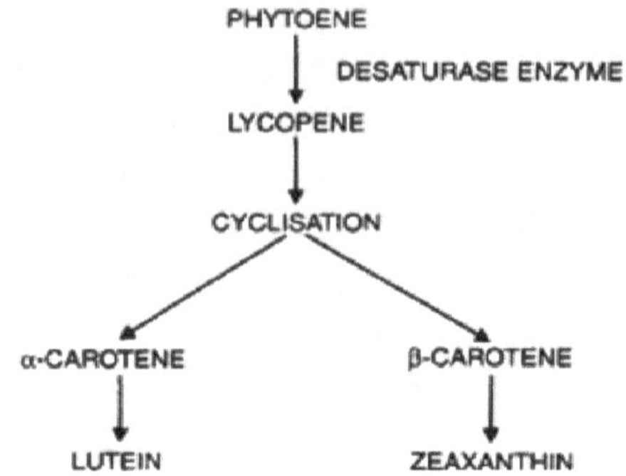

Fig. 2.6.4: Part of the biosynthetic pathway to carotenoids

Inhibition of Protoporphyrinogen Oxidase

There are several products that exert their effect through the accumulation of abnormally high levels of chlorophyll precursors. A structurally diverse range of herbicides has beenshown to inhibit the enzyme protoporphyrinogen oxidase, a pivotal enzyme at the branchingpoint of the porphyrin pathway leading to both haeme and chlorophyll biosynthesis. Theinhibitors of the process can be classiûed into three major chemical groups: the nitrodiphenylesters (actiûuorfen, lactofcn), the phenyl heterocycles (oxadiazon and sulfentrazone) and theheterocyclic phenylimides (flumiclorac).

Inhibition of Amino Acid Biosynthesis

Inhibition of Aromatic Amino Acid Biosynthesis

The shikimate pathway (Copping and Hewitt, 1998) is the biosynthetic source of threearomatic amino acids viz., phenyl alanine, tryptophane and tyrosine (Fig.2.6.5). These aminoacids are necessary for the protein synthesis as well as biosynthesis of auxin, most plantphenolic compounds and other secondary compounds.

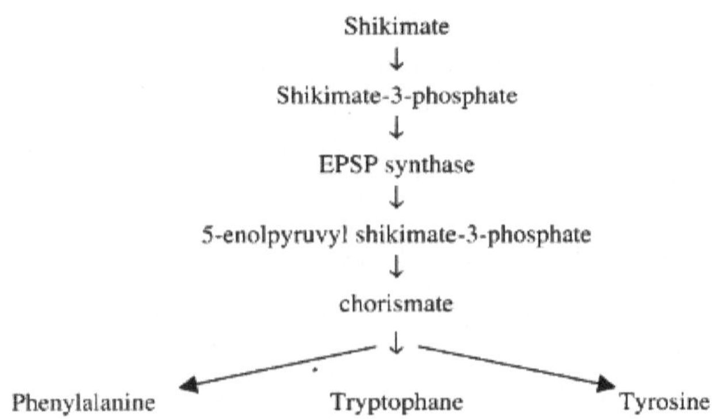

Fig. 2.6.5: Part of the biosynthetic pathway of aromatic amino acid biosynthesis

Branched Chain Amino Acid Biosynthesis

The branched chain amino acids: leucine, isoleucine and Valine are produced by similar biosynthetic pathways (Fig 2.6.6). In one pathway acetolactate is produced from pyruvate andin the other pathway acetyhydroxybutylate is produced from threonine. Both reactions are catalysed by the same enzyme that is known as both acetolactase synthase (ALS) andacetohydroxy acid synthase (AHAS).

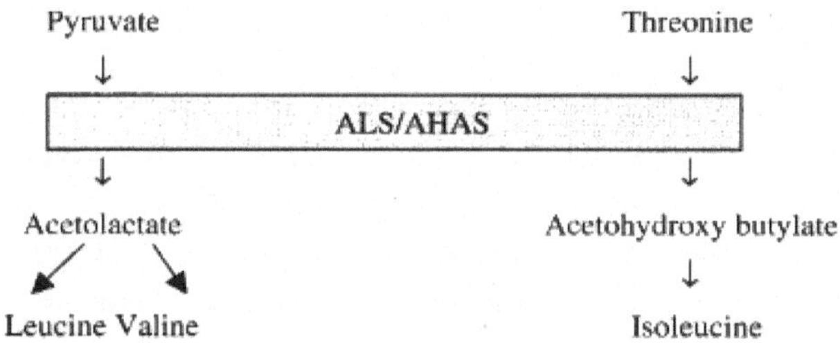

Fig. 2.6.6: Biosynthetic pathway to branched chain amino acids

Mitotic Inhibitors

Auxin Type Herbicides/Phenoxy Carbolic Acids

Despite many years of study, the actual molecular site of action of this family of herbicides is unknown. The Phenoxy carboxylic acids are sometimes called hormone typeherbicides because they mimic in many ways lethal doses of the plant hormone Indole aceticacid (IAA) (Devine *et al.*, 1993). IAA and PCAs are both thought to act by influencing plasma membrane properties by acting at a molecular site in the plasma membrane (Dekker and Duke, 1995). Initially these herbicides cause rapid changes in the elasticity of the plant cellswalls as a result of increased activity of plasma membrane AT Pases. Symptoms include epinastic bending with subsequent tumor-like unproliferated cell division at growing pointsand numerous secondary effects resulting in severe growth abnormalities and uncontrolled cell growth.

Dinitroanilines

Dinitroaniline herbicides affect cell division, elongation and differentiation (Holt *et al.*, 1993a). External symptoms include swelling of root tips and inhibition of elongation andlateral root development. Internal symptoms include formation of large isodiametricsometimes polynucleate cells in which mitosis is arrested. Cell wall and plates can bemalformed or absent. These symptoms are due to the effect of dintroaniline herbicides on tubulin (Hess and Bayer, 1977).

Carbamates

In general, carbamate herbicides are mitotic poisons. The soil active compounds check elongation and or expansion of plumules of germinating seedlings. Meristematic tissues basal to the first rudimentary leaf in the plumule are the most vulnerable to this effect (Banting, 1970). A shoot active carbamate like barban stops meristematic activity of axillary buds in wild oat. The affected seedlingshave been found to be low in serine, threonine, RNA, DNA and protein levels. The treated plants die because of prolonged static condition of growth.

N-arylalanine Esters

The N-arylalanine ester herbicides such as benzonylprop-ethyl and flamprop-methylprevent cell elongation in certain grass weeds allowing the crop to overtop them. The exact mode of action is not certain but it is proposed that these compounds interfere with the site ofaction of auxins (Copping and Hewitt, 1998).

Novel Mode of Action of Herbicides

The development of herbicides resistance within existing weed populations and shifts in weed population due to herbicide use and cultural practices have provided important stimulusto the search for new herbicides. Novel mode of action of herbicides offer additional tools tocombat the evolution of resistance to currently available herbicides and alternatives toherbicides to which resistance has already been developed. Herbicides compounds in thefuture should be more effective at lower rates, less toxic to other organisms, control resistantweeds and less prone to cause environmental problems. Recently, a new possible target sites for novel herbicide has been reported (Matsumoto, 1999). Some of these new targets are in the metabolic pathways that were not previously targets of herbicides. These findings indicate that the discovery of new target sites which are different with the available herbicides is highly practicable.

Herbicide Selectivity

Different plant species respond differently to a particular herbicide. Sometimes, even the plants of same species growing in different environments may also respond differently to the applied herbicide. When one herbicide is applied on variable plant populations, few plants arekilled and others are not affected at all. This phenomenon is called herbicide selectivity. In weed science, the meaning of the word 'selectivity' is different rather opposite when this term is used for crops and weeds. When an economical plant (crop) is not harmed by the applied herbicide that plant/crop is termed as selective to that particular herbicide whereas on the other hand the weed plants which are killed are called selective to the applied herbicide. The chemical which is safe to crop but phytotoxic to the target weed(s) plants is called selective herbicide.

Types of Selectivity

Selectivity is of two types:

Placement Selectivity

When the crop plants/crop seeds come in direct contact with herbicide, these cannot tolerate toxic effect but by putting the seeds below the treated zone or by avoiding crop exposure to herbicide, the harmful effect of these toxic chemicals on crop can be avoided andthis technique is called as placement selectivity. For example triallate (Avadex BW) is a wild oat specific herbicide and when it comes in direct contact with wheat seeds, it shows toxicityand when wheat seeds are placed below the treated zone, wheat crop becomes selective totriallate which is otherwise not.

Similarly any non-selective herbicide e.g. paraquat/glyphosate when sprayed as directed application only on weeds by using spray hood and protecting the crops particularly growing meristems from direct exposure to applied herbicide, then it acts as selective herbicide for the crop. In wider row sown crop, a non-selective herbicide can be made as selective herbicide bydirecting its application only on target weed plants. Out of these two contact herbicides i.e. paraquat and glyphosate, there is no significant loss of the crop if paraquat falls on the lowercrop leaves as it is non-translocated in nature and only the exposed plant portion will dry up.On the other hand glyphosate is a contact and translocated herbicide and when any plant part of the crop is exposed to this herbicide, the whole plant will be killed.

True Selectivity

When a mixed population of crop and weed plants, is exposed to an herbicide, weed plants are killed and crop plants are not killed and this phenomenon is called true selectivity. Thisselective kill of weeds may be due to genetic/ morphological or physiological differences. Rice and barnyard grass (*Echinochloa crusgalli*) are morphologically similar plants.Application of propanil herbicide kills barnyard grass but rice is not harmed which is due to the presence of arylacylamidase enzyme in the latter plant (crop) in higher concentrations as compared to former plant (weed). In rice, this enzyme hydrolysis the amide bond of propanil resulting into the formation of non-phytotoxic compounds i.e. 2, 4-dichloroaniline and propionic acid. This process is called inactivation of herbicide.

Factors on which Selectivity Depends

Selectivity/ sensitivity of a weed plant to a particular herbicide(s) depends upon many physical, biochemical and physiological factors. Most important parameters on which selectivity depends are as follows:

Stage/Age of Weed Plant

The frequency of weed kill with the applied herbicides is more when these are sprayed onactively growing stage. Uptake and translocation of the herbicide will be more by the actively growing weeds due to succulent tissues and more functioning of physiological processes. So, under such situations there will be more uptake of herbicide by roots/absorption by plant tissues, hence more translocation and plant will die quickly due to rapid movement of herbicide to the target site. In the other hand, sensitivity of weed plants to herbicide decreases with age due to dilution effect and hardiness of the tissues. In older plants, rate of movement of metabolites with which herbicide move in the plant system would be low as compared to young plants (actively growing) which (older plants) may escape because lethal concentration of herbicide will not be maintained.

Morphological Differences

Morphology of weed plant also plays very important role in selectivity. In general broad leaf weeds due to more leaf area and more appropriate leaf angle, receive (intercept) more amount of applied herbicide and are more susceptible than grass type weeds, whose position with stem is vertical during initial stages which intercept less herbicide. Secondly, weed plants with smooth leaf surface retain less spray fluid as compared to weed plants with rough leaf surface. However, presence of small and thin pubescence helps in improving action ofherbicides. On the other hand, a weed plant with very dense cover of hairs escape from the toxic effect of herbicide as cuticular layer will either not receive spray droplets or will receive less spray fluid and thereby lethal concentration will not be achieved.

Physiological Factors

The fundamental principle of herbicide selectivity is that more toxicant (herbicide reaches at the site of action in an active form inside the target plants (weeds) than in the non target species. This may occur due to: 1. Differential absorption, 2. Differential translocation, 3. Differential deactivation of the applied herbicide and 4. Differential metabolic processes.

Differential Absorption of Herbicide

Absorption is the movement of herbicide or other material from the surface into the plant system in a limited zone. Foliar applied herbicides are absorbed mostly through leaves,whereas, soil applied herbicides through roots.

Absorption of Herbicide from Soil

Many herbicides kill the germinating seedlings, so these are applied either as pre-emergence or as pre plant soil incorporation. The absorption of herbicides applied to soil is usually associated with roots of the plants. Herbicides may also be absorbed by the swelling seeds and coleoptile, epicotyl and hypocotyl of the emerging seedlings. The shoot zone of germinating plants in soil (coleoptile etc.) has been revealed as a major site of absorption of many soil applied herbicides (Barrentine and Warren, 1971). Prendeville *et al.* (1968) found that EPTC, chlorpropham and trifluralin were most effectively absorbed when applied to the subsurface shoot zone of the test plant namely maize and peas. But 2, 4-D, simazine and naptalam etc. were absorbed mainly from the root zone. Entry of chlorthal methyl in sorghum was primarily through its shoot zone in soil.

1. **Interception**: When roots of plants (weeds) penetrate in soil, they come in direct contact with the herbicide molecules.
2. **Mass flow**: It is flow of herbicide molecules with aqueous soil solution. Massflow is passive process whereby the herbicide molecules are transported in soil with the soil solution to the seeds, roots or underground shoots and later on into other plant parts.
3. **Diffusion**: It is a passive process in which the herbicide molecules move along a concentration gradient in the soil solution towards roots of plants. It also involvesthe physical process of diffusion of substances through the plasma-membranes andis controlled by the electrochemical potential gradient across the plasmalemma and tonoplast of the cytoplasm.

Foliar Absorption

The absorption of herbicides involves their entry from plant surfaces into the cytoplasm of epidermal cells of shoots. In this process herbicide may come across many possible barriers which are:

1. Waxes and Hair

The outermost coverings of many plant foliage that are exposed to the atmosphere arecovered with waxes and hairs. Plant surface exposed to full sunlight develop thicker waxcoatings than those exposed to partial sun light. Waxes and hairs are lipophilic in nature and they interfere in wetting of leaf surface with spray fluid which is lipophobic in nature. Molecules or surfaces that are attracted by oil-like surfaces and molecules, but are repelled by water are called Lipophilic (= hydrophobic). On the other hand molecules or surfaces that are attracted towards water, but repelled by oil-like surfaces or substances are called Lipophobic (= hydrophilic).

2. Cuticle

Cuticle is an uneven lipoidal covering on the plant leaves and branches to guard against excessive transpirational loss of water from plants. The cuticle is semi-lipoidal, oxidative polymer of 18 or more carbon hydroxyl fatty acids and alcohols. Thus, cuticle carries both hydrophilic and lipophilic groups which impart to it a semihydrophilic character. Cuticle is permeable to both hydrophilic (polar) and lipophilic (apolar) substances. The roots of plants have no cuticle; therefore they absorb hydrophilic substances more readily than the lipophilic compounds.

3. Stomata

Stomatal cavities increase the absorptive area of leaves. Herbicide spray drops that may enter them, pass through their thin, internal cuticle more rapidly than through the thick surface cuticle.

4. Cell Wall

After crossing the cuticle, herbicide come in contact with cell wall of epidermal cells. The primary cell walls are rich in pectins and celluloses, hence, they are more hydrophilic than the cuticle.

5. Plasmalemma

Plasmalemma is a lipoidal, differentially permeable membrane lining the plant cells just below the cell wall. It is the last barrier to the entry of herbicide into the cytoplasm of epidermal cells.

Absorption of herbicides in tolerant plants is so slow that it kept pace with the metabolism easily, thus the plant proved tolerant.

Differential Translocation

Herbicides are absorbed either by roots or foliage or both depending upon the type andnature of herbicide. Their site of action is different and these herbicides are expected to reachat target site before they show toxic effect. Transport of herbicides is of two types:

1. Acropetal (through xylem vessels), 2. Basipetal (through phloem vessels)

Movement of root uptake herbicides in plant is through xylem vessels along with transpiration stream, whereas of foliar applied herbicides is through phloem with photosynthates. Rate of translocation of herbicides depends upon rate of weed growth and nature of herbicide. For instance, under favourable conditions 2, 4-D may translocate through the phloem as fast as 160 cm per hour. When phloem tubes are damaged due to some reasons, even a normally translocated herbicide will fail to translocate in plants. Concentration of herbicide also governs speed and distance of translocation. For example, when dalapon is applied at low concentrations of 0.2 to 0.5 percent to a healthy grass clump, it gets translocated easily to the roots. But when spray concentration is high (1-2%), dalapon will quickly brown the treated foliage without translocating appreciably to the grass roots. Likewise, paraquat, diquat, glyphosate herbicides are immobile in plants mainly because of their rapid phytotoxicity to leaf tissue even at low concentration. Selectivity of a crop plant to a particular herbicide sometimes also depends upon rate of its translocation in the plant system.

Types of Translocated Herbicides

The translocation behaviour of herbicides differ with the herbicide structure and sometimes it may alter with the plant species, environment, etc. Depending upon these factors, five important types of translocated herbicides are recognised (Gupta, 1998).

1. Shoot Mobile Herbicides

These are foliar applied herbicides that show both acropetal and basipetal translocation through phloem, sometimes simultaneously. They translocate both to shoots and roots of thetreated plants. Examples, amitrol, glyphosate, dalapon, dicamba, TBA etc.

2. Shoot Immobile Herbicides

They are foliar applied herbicides that translocate only in the acropetal direction through xylem. They do not translocate from leaves to the roots of the treated plants. Exampleatrazine and diuron.

3. Circulatory Herbicides

These are foliar applied herbicides capable of frequent phloem-xylem interchange.

Examples MH, 2, 3-6 TBA, picloram, dalapon and amitrol.

4. Reverse Xylem Mobile Herbicides

They are foliar applied herbicides that translocate to roots through xylem in acute Water deficiency conditions. Examples: arsenic trioxide ions and sodium chlorate. Paraquat when applied to leaf in subduded light takes more time for desiccation as compared to its application in bright sun light conditions.

5. Root Mobile Herbicides

These are soil applied herbicides and translocate freely in xylem tissues to reach the plant tops. Examples, dicamba, tricamba, uracils, S-triazines and substituted ureas.

Phytotoxicity

Phytotoxicity refers to the injury caused by a agro-chemical on a plant.

Herbicide Toxicity Ratings

The crop toxicity ratings were made on zero to ten point scale for first few days after application of herbicides. The ratings are as follows :

Effect	Rating	Effect on crop
None	0	No injury, normal
Slight	1	Slight stunting injury or discolouration
	2	Some loss of stand, stunting or discolouration
	3	Injury more pronounced, but not persistent
Moderate	4	Moderate injury, recovery possible
	5	Injury more persistent, recovery doubtful
	6	Near severe injury, no recovery possible
Severe	7	Severe injury stand loss
	8	Almost destroyed, a few plants surviving
	9	Very few plants alive
	10	Complete destruction

Weed Control Ratings

The qualitative assessment of weed control by herbicides would be made through visual ratings. The ratings can be made on zero to ten points scale for first few days after herbicide application. The ratings are as follows.

Effect	Rating	Effect on crop
None	0	No control
	1	Very poor control
	2	Poor control
Moderate	3	Poor to deficient control
	4	Deficient control
	5	Deficient to moderate control
	6	Moderate control
Severe	7	Satisfactory control
	8	Good control
	9	Good to excellent control
	10	Complete control

The visual symptoms of herbicidal injury are caused by the deficiency of certain metabolites or substances required for the growth of plants. They are described below:

1. Wilting

Wilting and desiccation of plants are common symptoms of membrane disruption resulting from diffusion or leakage of solutes from cells to the intercellular spaces followed by plasmolysis. Metsulfuron-methyl, 2, 4-D, imazethapyr, tribenuron-methyl induces wilting in plants.

2. Chlorosis Followed by Necrosis

Leaf chlorosis followed by necrosis is a common response of plants on exposure to the photosynthesis-inhibitor herbicides. Almost more than 50% of the herbicides used in agriculture are photosynthesis inhibitor directly or indirectly. They include herbicide groups such as, phenylureas, uracils, triazines and herbicides like propanil, pyrazon, bentazon, sulfo sulfuron.

3. Necrosis

Here only necrosis occurs and no chlorosis. Paraquat, diquat, bromoxynil, ioxynil, nitrofen, endothall, fluorodifen, cacodylic acid and acrolein induce necrosis. They all are contact herbicides and cause disruption of cellular membrane in plants.

4. Leaf Burning

Paraquat and diquat cause burning of leaves by rapid desiccation. Even isoproturon produces this symptom on *Avena ludoviciana* (wild oat) and *Phalaris minor* (little seed canary grass). The leaves start burning from the top/apex.

5. Stunted or Retarded Growth

Growth of plants becomes very much stunted as a result of inhibition of cell elongation. Chlorsulfuron, benzoylprop-ethyl, flamprop-isopropyl, flamprop-methyl and sulfosulfuron causes retarded growth of grass weeds, namely *Avena ludoviciana* and *Phalaris minor* in winter cereals mainly wheat.

6. Epinasty/epinastic Symptoms

Epinasty means bending, twisting or curling of leaves, stems or even ear head, which later become brittle and break up even by light pressure. Epinastic effects are usually associated with altered cell division, cell elongation and tissue differentiation, *e.g.,* 2, 4-D, MCPA, MCPB (Phenoxylalkanoic acids), 2,3, 6-TBA, dicamba (Benzoic acids), fenac (Phenyl alkanic acids) and picloram (Picolinic or Pyridine carboxylic acids). They may also cause adventitious roots, callus or fascinated organ formation.

7. Formative Effects

Formative effects like epinastic effects are also associated with the altered cell division, cell elongation and tissue differentiation. Chloro-acetamides, dinitro aniline, nitriles, thio carbomates, barban, bensulide, chlorpropham, dalapon, DCPA, diphenamid, glyphosate, naptalam, phenoxyalkanoic acids, propham, pronamide, TCA produces formative effects in plants.

a) Increased tillering due to apical portion damaged

b) Multiple shoot formation at one node

c) Shorter internode, thickened leaves, coleoptile and leaf distortion

d) Abnormal seedling elongation

e) Reduced secondary root formation

f) Albinism of leaf occurs following treatment with amitrole, flouridone, norflurazone, clomazone (in wheat) on newly developing leaf. Chloroplast formation is affected

g) Earhead malformation: 2, 4-D causes earhead malformation in some varieties of wheat *e.g.,*Arjun (HD 2009), Kundan. Of course, the percentage of occurrence of malformed earhead is very less. The malformed earhead is not upright or straight, usually bent and twisted having a bit disrupted look.

Crops

All the recommended herbicides are safe for respective crops in which they are recommended provided they are used according to the recommended technology. Otherwise, crops may show toxicity symptoms under certain situation such as :

- **Wrong time of application:** Application of herbicides too early than its recommended time may cause yellowing and necrosis of crop. Over dose of 2, 4-D and its application on very young crop may cause tubular leaves and deformed wheat ears.
- **Use of over dose:** Application of higher than recommended dose is also responsible for stunting, yellowing and necrotic growth of crop plants
- **Wrong application technology:** Un-even application of herbicides not only harms thepresent crop but may also prove harmful for the succeeding crops due to residual toxicity.
- **Spray drift:** Due to speedy winds on the day of spray, adjoining crops may be damaged.
- **Adverse climatic conditions:** Due to heavy rains immediately after the application of herbicides, few root uptake herbicides may even damage the tolerant crops.
- **Wrong herbicide used:** With the use of unapproved herbicides, crop is damaged. For instance, wheat crop may be damaged with the use of metribuzin. Germination of wheat crop is adversely affected when seeds come in contact with triallate treated soil zone.

Weeds

Toxicity is adverse effect of herbicide on the growth and development of plants.

Type of Toxicity

It is of two types

1. **Acute Toxicity:** When the effect of herbicide is not having quick effect and shows its phytotoxic treatment. *e.g.,* Effect of gramoxone
2. **Cronic toxicity:** When a herbicide is not having quick effect and shows its phyto-toxic effectafter few days of application *e.g.,* Effect of isoproturon on *Phalaris minor.*

Deactivation of Applied Herbicides

Herbicide selectivity is primarily a function of differential rates of deactivation of thetoxicant by the target and the non target plants before they reach their respective site of action. This deactivation may be a process of: I. metabolism, II. reverse metabolism, III. conjugation, IV. accumulation and V. secretion. A tolerant plant species deactivates theherbicide molecule (or ion) rapidly, where as a susceptible species does so, slowly.

Metabolism

Metabolism involves a change in the molecular structure of the applied herbicide inside plants to yield some non-phytotoxic products. This process is more rapid in tolerant plants.The tolerance of rice to bensulfuron and Lolium sp. to chlorsulfuron was due to their rapidmetabolism inside plants.

Reverse metabolism

Sometimes the metabolites produced in the plant due to some metabolic reactions aremore phytotoxic than the parent compound and this process is called reverse metabolism. The herbicides 2,4-DB and MCPB, per se are non-phytotoxic but in sensitive plants with enzymicB-oxidation process, they are converted to toxic chemicals i.e., 2,4-D and MCPA respectively. This reaction occurred in many non-leguminous plants and certain legumes like pea, lucerne, berseem, etc. either lack the necessary B-oxidation system or operated its luggishly and consequently these legumes proved comparatively tolerant to 2,4-DB and MCPB. It has also been reported that metabolites of sulfosulfuron herbicide after its application have more residual effects on succeeding crops than pure chemical i.e.sulfosulfuron.

Conjugation

Conjugation is coupling of intact herbicide molecules with some plant cell constituent in living plants. Stoller (1970) found that ^{14}C chloramben conjugated rapidly inside the roots of soybean (*Glycine max*) with glucose molecules, forming N-glucosylchloramben and an unknown compound chloroamiben-X.

Accumulation

Accumulation is another method of deactivation of herbicide and its translocation in the tolerant plants is checked. Prometryn is accumulated in the lysigenous glands in cotton, atolerant plant whereas it is distributed throughout in soybean, a susceptible plant.

Secretion

Some plant species are able to secrete phloem transported herbicides from their roots e.g. secretion of 2, 4-D from Jimson weed.

Metabolic Processes

It is one of the most important basis of herbicide selectivity. For growth and development of plants, many metabolic reactions/changes goes on in the plant. Herbicide interferes with these metabolic processes going on in plants and these are being discussed group/herbicide wise:

S-triazine

Simazine has been known to be metabolised in resistant maize to hydroxy simazine with the help of a compound identified as 2,4-dihydroxy-7-methoxy-1, 4-benzoxazin-3-one (MBOA) present in roots of such plants. The hydrolysis was found to be rapid enough to prevent the entry of intact molecules of absorbed simazine into the leaf chloroplasts of maize.

Substituted Ureas

Dealkylation of substituted ureas is common in the tolerant species of plants Frear *et al.* (1969) isolated from leaves of cotton, broadbean and buck wheat - all tolerant –amicrosomal, mixed function oxidase enzyme which has found to catalyze N-demethylation of 3-(phenyl)-l-methylurea herbicides e.g. monuron, diuron and fluometuron. Each mole ofsubstrate that was demethylated produced l mole each of the formaldehyde (HCHO) and –chlorphenylurea.

The selectivity of wheat to isoproturon is based on its degradation throughmonoxy generation. These degradations are performed by membrane-bound NADPH-dependentcy lochrome P-450 monoxygenases.

2.4-D

The tolerant plants have been found to metabolize 2, 4-D molecules by side chain degradation, followed by ring hydroxylation. The side chain degradation involves decarboxylation, demethylation and cleavage of the ether linkage.

Propanil

Selectivity of propanil between resistant rice (*Oryzasativa*) and susceptible *Echinochloa colonum* biotypes was found to be due to differences in the arylacylamidase enzyme contents of their leaves. The enzyme is capable of hydrolysing propanil in plants to non-phytotoxiccompounds, 3-4-dichloroaniline and propionic acid. Frear and Still (1968) found that susceptible *E. colonum* leaves contained only l/60th the quantity of the enzyme present in therice leaves. In rice too, this enzyme level could fall greatly in response to either an abnormally high atmospheric temperature or application of certain insecticides with propanil.

Diclofop

In tolerant wheat diclofop is aryl-hydroxylated and conjugated to an aryl glucoside. In susceptible wild oat the conjugate is neutral glucosyl ester which is easily reversible to toxicdiclofop. In addition, rates of herbicide adsorption and metabolism may play important rolewith some plant species (Boldt and Putnam, 1981).

Metolachlor

Selectivity of metolachlor between maize and yellow nutsedge was found due to chieûydifferential metabolism of the herbicide. In addition, yellow nutsedge shoots readilyabsorbed the herbicide, while in maize it was slow (Dixon and Stoller, 1982).

Ethametsulfuron-methyl

In tolerant plants, like the rape (*Brassica napus*), this herbicide was found rapidly metabolised. The major metabolites were O-desmethyl and N-desmethyl-O-desmethyl of theherbicide molecule; both being non-phytotoxic compounds (Lichtner, 1995).

Absorption and Translocation of Herbicides in Crops

To prove phytotoxicity to any plant, a herbicide must reach appropriate sites within the plant at which an ultimately lethal chain of events occurs. Thus successful herbicide userequires more retention on plant surfaces, entry of the chemical into the plant, movement by diffusion or other means through the plant, escape from detoxification mechanism, and finally an attack at the molecular level upon some process vital to plant survival. Sometimes, lethal dose of herbicide do not reach to the site of action in the target plant due to hindrancein absorption or translocation of the applied herbicide. So, these plants escape to the phyto-toxic effect of the chemical. Some herbicides may undergo decomposition processes or someinactive conjugates are produced by which target plant escapes.

Plant Absorption of Herbicide

Adequate absorption of post-emergence herbicides by the plant (weed) canopy is very important. Due to bigger spray droplets, there is more bounce-off of the herbicide solution. So retention and consequently penetration and translocation of applied herbicide in the target plantswill be less. Absorption of herbicide depends upon following factors:

l. Leaf area

If leaf area is more, then more herbicide will retain on the foliage. Thus more penetration and more translocation of chemical are expected. An adequate leaf area at the time of spray is essential for herbicides whose uptake is only through leaves and not through roots e.g.clodinafop. The portion of the herbicide which is not received by the foliage goes waste. So, foliar acting herbicides should be applied when most of the weed seeds have germinated and they are in active growing stage.

2. Leaf Structure

Absorption of herbicide also depends upon nature of leaf structure which may be smooth or rough. There is more absorption of applied herbicide solution by the weeds having rough leaf structure than the weeds having waxy layer on the leaves (smooth). Presence of leaf hairs also govern absorption of applied herbicide. Dense hair cover will not allow herbicide solution to come in contact with leaf surface (cuticle) and hence less absorption. On the other hand presence of small and thin pubescence helps in improving action of herbicide by enhancing its absorption.

3. Leaf Angle

Erect/vertical leaf canopy intercept less applied herbicide as compared to the canopy with drooping or horizontal leaves. Bouncing off of spray droplets on ground surface will be more in vertical crop canopy and less in horizontal crop canopy. In general, the angle of broad leaf weeds is more appropriate for effective herbicide retention and hence, more absorption than that of grass weeds.

4. Growing Point

In monocot weeds growing point is enclosed and in dicot weeds it is exposed. Dicotweeds are more prone to toxic effect of herbicide as site of action of most of the herbicides isin growing meristems. Toxicity in grass weeds is comparatively less due to protects;growing points as compared to broad leaf weeds.

5. Type of stem

Hairy and grooved stems retain more quantity of herbicide than smooth/glazed sternStems having thin bark show more absorption than stem having thick bark. Apart from thesepresence of small spines on the stem also help in more retention and absorption of herbicide.

Mechanism of Herbicide Absorption

Absorption Through Cuticle and Cell Walls

Hydrophilic ions and molecules enter plants by the process of diffusion, whereas the lipophilic substances enter by their lipid solubilizing action. Molecules or surfaces that are attracted by oil-like surfaces and molecules, but are repelled by water are called Lipophilic orHydrophobic. On the other hand Lipophobic (= hydrophilic) molecules or surfaces that areattracted towards water, such as sodium chloride. They are repelled by oil-like surfaces orsubstances. The lipophilic spray liquids solubilize the native lipid molecules of cuticle and epidermal walls and increase their permeability to the herbicide. Solubilisation is a moreactive process of absorption than diffusion. Many adjuvants impart certain lipophilicproperties to the aqueous spray liquids and aid entry of hydrophilic substances through thecuticle and cell wall. Lowering of pH of spray liquids of acid type herbicides, such as 2,4-Dand silvex, also exerts an absorption enhancing effect by limiting ionization of their molecules. Herbicides are absorbed through cuticle more rapidly in molecular forms than intheir ionic form. Herbicide absorption through cuticle and epidermal walls is reversible. Aheavy rain can wash out at least a portion of the absorbed herbicide, if it was still outside theplasmalemma at the time of rain.

Absorption Through Plasma Membrane

The plasma membranes comprising plasmalemma and tonoplast are made up chiefly of lipids and proteins and are lipophilic in nature. The lipophilic herbicides such as herbicidal oils and oil emulsions pass through them easily at rates roughly proportional to their fatsolubility and they reach the epidermal cytoplasm rapidly. There are two probable processesby which hydrophilic herbicides may find their entry into epidermal protoplasm through plasma lemma.

1. **The passive absorption :** This involves the physical process of diffusion and is controlled by the electrochemical potential gradient across the plasmalemma and tonoplast ofthe cytoplasm.
2. **The active process :** This is an energy requiring mechanism and is based on the hypothesis that certain heavy chemical carrier molecules, constituting primary proteins and phospholipids and possessing both positive and negative ends, occur within the plasma lemma. These chemical carrier molecules are instrumental in transporting anions, cations and undissociated molecules of specific solutes across the plasmalemma into the cytoplasm. The tonoplast or vacuolar membrane too has a mechanism similar to the plasma lemma for transporting the absorbed solute from cytoplasm into the vacuole. But the tono plast carrier system is comparatively very weak.

In soil contact between herbicide and plant roots is mainly due to following processes

1. Interception

When roots of plants grow/penetrate in soil, they come in immediate contact withherbicide. This process is called interception.

2. Mass flow

It is a passive process whereby the herbicide molecules are transported along with the soil solution to the seed, root or underground shoots by the flow of this aqueous solution towardsor into these plant parts.

3. Diffusion

It is again passive process in which the herbicide molecules move along a concentrationgradient in the soil solution towards roots of the plant.

Movement from Site of Uptake to Site of Action

Herbicides applied to foliage can be divided into two groups i.e., those that primarilyaffect only sprayed foliage (contact herbicides) and those that are translocated to other partsof the plant as well (translocated herbicides). Contact type of herbicides do not move in theplant and they are unable to move in the plant system e.g. paraquat. Also these herbicides arecalled non-translocated herbicides.

Translocation of Herbicide

After the herbicide ions and molecules have moved to the cells adjacent to the vascular vessels i.e. xylem and phloem tubes, they must move into appropriate conducting tissues fortheir translocation to distinct plant parts. Generally these are three types of translocation.

1. Intracellular: It is the movement of herbicide ions and molecules within an individual cell and it involves short-distance transport.
2. Intercellular: It is movement of herbicide ions and molecules from cell to cell via the symplast (short-distance transport) and the xylem and phloem conducts (longdistance transport). Generally non-polar substances with low interfacial tension may move through the inter-cellular spaces.
3. Extracellular: Is the movement of herbicide ions and molecules through cell wall and intercellular spaces within the apoplast and it primarily involves short-distancetransport.

2.7

Herbicide Mixtures, Combinations, Rotations and Interactions

Herbicide Mixtures

A mixture of two or more herbicides could be more effective for controlling different weed species. In horticultural and plantation crops, weed free situation is needed during the entire year over several years. In such situations, herbicides are to be rotated to take care of all the species of weeds. A mixture of herbicides, due to chemical reaction, may enhance or reduce the efûciency of one or more compounds. Hence, knowledge of herbicides interactions is essential for efûcient use of herbicides in weed control.

Herbicide Combinations

Herbicide combinations or mixtures are chemicals containing two or more herbicides for effective and economical weed control. They offer several advantages:

- A mixture will broaden the spectrum of herbicidal action for controlling different plant species,
- Synergistic or additive effect due to combinations may increase weed control efûciency,
- Component herbicides in a mixture may prevent rapid detoxification or degradation of the other leading to its efficiency, and
- Scope for reducing the rate of application of each of the herbicides required for weed control.

There are, however, limitations in the usage of herbicide combinations.

- Combination of two selective herbicides may become non-selective.
- Antagonistic reaction of two herbicides will lower its efficiency relative to either of them applied independently, and
- Physical compatibility may be a problem in some herbicide combinations.

There are two types of herbicide mixtures: tank mixtures made with desired herbicides and rates before application and the concentrated herbicide mixtures formulated by the manufacturers. Several herbicide combinations are available under various trade names to suit speciûc weed problems. Many of the combinations involve triazine, urea, uracil and phenoxy acid herbicides.

Herbicide Rotations

In annual cropping systems, weed free environment is necessary during the entire year and in plantation crops for over several years. Under such situations, use of the same herbicides several times or different herbicides of herbicidal combinations is needed. Soil applied or foliage applied herbicide or both are used in sequence to take care of annual as well as perennial weeds. Herbicide rotation offers several advantages.

- Effective weed control for the crop duration,
- Preventation of emergence and occurrence of tolerant weed species,
- High cumulative cost: beneût ratio over years,
- Reduction in the rate of application of herbicides over years, and
- Low levels of herbicide residue build up.

Herbicide Interactions

When two or more herbicides are applied in a mixture or alone as sequential applications, they interact leading to synergistic, additive or antagonistic responses. An herbicide may show synergistic response in combination with one herbicide and antagonistic response with another herbicide.

Synergistic and Additive Responses

Synergistic responses in herbicide combinations are related to 2, 4-D and other phenoxy acid derivatives. A mixture of 2, 4-D and chloropham show synergistic effect on monocot species, generally, resistant to 2, 4-D. Low rate of 2, 4-D and picloram show synergistic response on *Convolvulus arvensis*. Combination of 2, 4-D, mecoprop and dicamba have synergism on several dicot weed species.

Combination of 2, 4-D and triazine herbicides also shows synergism. Combination of 2, 4D, atrazine and crop oil is much more effective that: atrazine cooled with oil. Amitrole and ammonium thiocyanate combination: is more effective than ammonium thiocyanate. *Agropyron repens* can be effectively controlled by combination of paraquat, triazines and ureas. Paraquat and Pentanochlor combination also gave synergistic effect on weedcontrol. Atrazine and alachlor combination is widely used for effective weed control in maize.

S.No.	Herbicide combination	Trade name
1.	Atrazine + Propachlor	Astram
2.	Atrazine + Prometone	Astratol 80W
3.	Atrazine + Simazine	Pintax
4.	Atrazine + Amitrole + Finac	Fenamine
5.	Atrazine + Chlorate + Borate	Atratol 80W
6.	Borate + Bromacil	Borasen
7.	Borate + Diuron	Monobor-chlorate
8.	Borate + Monuron	Ureabor
9.	Borate + Sodium chloride	Chlorax-40
10.	Borate + 2,3,6-TBA	Benzabor
11.	CDAA + TCBC	Randox T
12.	Dalapon + TCA	Dowpon
13.	Diuron + Bromoxuron	Brute
14.	Diuron + Bromacil	Krovar
15.	2, 4-D + Borax	DB granular
16.	2, 4-D + Fenac	Fenac plus
17.	2, 4-D + Dicamba	Benevel, Brush Buster, Banvel-K,
18.	2, 4-D + Dichlorprop	Weedone Super D
19.	2, 4-D + EPTC	Bush killer 170
20.	2, 4-D + MCPP	Knox weed
21.	2, 4-D + Mecoprop	Chipco, truf Kleen
22.	2, 4-D + MSMA	Mecopar
23.	2, 4-D + Picloram	Mad
24.	2, 4-D + 2, 4-DP	Amadon 101, Tordon 101, Tordon 212
25.	2, 4-D + 2, 4, 5-T	Envert-171, Weedone-170
		Amine 2D/2T, ded-weed LV33,
26.	2, 4-D + 2,4,5-T + Dicamba	Dinoxol, Envert171,
27.	2, 4-D + DSMA + Bromacil	Easteron, Emulsamine BK,
28.	2, 4-D + Sufactant	Emulsavert 248,Lo-Vol
29.	Endothall + Silvex	2D/2TBrushkiller-64, Weeder-
30.	Endothall + Ammonium sulphate	Amine-BK 64
31.	Linuron + Amitrole	Banvel, Brush Buster -III
32.	Linuron + Chloramben	Wood broom
33.	Linuron + Propachlor	Weed-E-Red 360,Ansar,DSMA liquid
34.	MCPA + Bromoxynil	Aquathol plus
35.	MCPA + Dicamba	Des-i-cate sulphate
36.	MSMA + Surfactant	Kleer-Lot
37.	MSMA + Cacodylic acid	Amilon
38.	Naptalam + Chlorpropham	Londax
39.	Naptalam + Dinitro	Brominal plus, Bronate
40.	Triflualin + Diphenamid	Mon Dak
		Ansar 529 HG, Weed Hoe-108
		Chex-mate
		Alanap plus, Mor-com, Solo
		Kleenkrop
		Trefmid

Antagonistic Response

Combination of EPTC with 2, 4-D, 2, 4,5-T or dicamba gives antagonistic response in sorghum. Simazine or atrazine added to glyphosate reduce glyphosate activity. Generally, contact and systemic herbicide combinations show antagonistic responses. Tank mix application of paraquat with glyphosate or dalapon leads to poor control of *Imperata cylindrica*. Several foliage applied herbicides used for wild oat control lose part of their activity when applied as tank mixture with auxin herbicides such as 2, 4-D, MCPAand dicamba.

Herbicide-Antidote Interactions

Herbicide-antidote interactions are antagonistic in nature. Antidotes like R25788 and CDAA reduce toxicity of alachlor, EPTC and butylate to certain plants. Seed treatment with NA protects maize seedlings from EPTC injury. Coating oat seed with NA reduces the phytotoxicity of barban. Activated charcoal serves as an effective barrier to the availability of soil applied herbicide to the plant and reduces toxicity hazard of the chemical to a desired plant.

Herbicide-Insecticide Interactions

Tolerance of plants to a herbicide may be altered in the presence of an insecticide and vice versa. Phytotoxicity of monuron and diuron cotton increases when applied with phorate. Phorate or disulfoton with metribuzin decreases soybean yield. Combination of alachlor and carbofuran leads to reduction in barley yield. Propanil after Carbofuron treatment increase phytotoxicity on rice. In general, risk of crop injury, stand reduction and decreased yields could result when combinations or sequential applications of herbicides and insecticides are used.

Herbicide-Fertilizer Interactions

Addition of ammonium nitrate solution (0.1%) increases the absorption of isooctyl ester of 2, 4, 5-T for its effective use. *Imperata cylindrica* can be effectively controlled by addition of ammonium sulphate at 0.5 per cent (W/V) concentration to glyphosphate. In general, many herbicides are equally effective whether applied separately or in combination with fertilizers. All herbicides show chemical and physical compatibility with suspension fertilizers.

2.8

Herbicide Persistence and Residue Management

Persistence of Herbicide in Soils

Most of the herbicides are applied to soil as pre-emergence for weed control under different situations. For effective weed control, herbicides must remain in soil in active and available form until their purpose is served. Larger persistence poses hazard to subsequent land use and is undesirable. The length of time that an herbicide remains active in the soil is called persistence or soil residual life. The quantity of herbicide that remains in the soil after its mission is accomplished is referred as residue. The term residue infers the continued presence of herbicide in the soil in very small quantities even though it may not be present in herbicidal quantities or available for plant absorption. Herbicide residues in soils are undesirable mainly because they may

- Injure sensitive crops in a cropping system,
- Be absorbed by succeeding crops and residues accumulated in unlawful amounts in the produce,
- Result in building up of herbicide residues following repeated treatments that exceed their rate of dissipation, and
- Inhibit growth of soil micro-organisms.

Knowledge of persistence and carry over residual effect of herbicides is essential to use them safely and effectively. Persistence of herbicides in soils pose hazard when

- Highly persistence herbicides are used,
- Crop failure necessitates replanting,
- A susceptible crop follows a short term crop which received a persistence herbicide, and
- Decomposition of the applied herbicide proceeds very slow.

Fate of Herbicides in Soil

The fate of herbicides in soils pertains to their interaction with soil environment. Factors inûuencing persistence of herbicides are:

- Microbial processes,
- Chemical decomposition,
- Photodecomposition,
- Adsorption by soil colloids,
- Plant uptake,
- Volatilization,
- Leaching, and
- Surface runoff.

Microbial Processes

Soil microbes bacteria, fungi, actenomycetes and algae may interact with herbicides. Microbial decomposition (biodegradation) is the major mode of decomposition of herbicides in soils. Herbicides belong to the group sphenoxies, phenylureas, aliphatic acids, acetanilides and carbonates are readily attacked by a host of soil microbes. Microbes degrade herbicides predominately through hydrolysis, alkylation, oxidation, reduction, hydroxylation, ring cleavage and conjugation.

Chemical Decomposition

Chemical decomposition involves such processes as oxidation, reduction, decarboxylation, hydroxylation, hydrolysis, dealkylation, conjugation and ring cleavage. Oxidation-reduction reactions involve transfer of electron from one reactant to another resulting in the formation of electrically charged (ionized) compounds. Herbicide molecules having ring structure such as derivatives of phenoxy, benzoic and triazine groups are decomposed by hydroxylation. Phenoxy esters, carbamates, triazines and ureas are rendered inactive by hydrolysis.

Adsorption by Clay Colloids

All soil applied herbicides are adsorbed to some extent and their activity is reduced in direct proportion to the amount adsorbed. Adsorbed herbicide molecules are in passive state, unavailable to biological, physical and chemical processes until disorption occurs. In general, herbicides are more strongly adsorbed in dry soils than in wet soils and similarly in humus than in clay. In particular, cationic herbicides are more strongly adsorbed to clay particles than to humus, anionic herbicides more strongly to clays of montmorillonite type and less strongly to kaolinite and hydrous mica clays. Herbicides are adsorbed more strongly in acid soils than in alkaline soils.

Photodecomposition

Some organic herbicides undergo molecular alteration on exposure to sunlight due to photochemical reaction leading to their degradation. Under field conditions, photodecomposition is an important mechanism for deactivation of herbicides exposed on soil surface. Incorporation of herbicides can reduce such decomposition. Herbicides such as amiben, triafluralin, certain substituted urea and triazinesare susceptible to photodecomposition.

Plant Uptake

Herbicides are absorbed and metabolized by plants. They return to soil either in the original form or as metabolites through crop and weed residues. In the case of fenac and 2, 3, 6-TBA, a portion of absorbed herbicide may be circulated through the plant and returned to the soil in an unchanged form by root exudation or a part of herbicide absorbed by plants is retained in the edible sink of the plant and the rest is returned to the soil through crop residues.

Volatilization

In this process, a chemical passes from solid or liquid state to the gaseous state. Herbicides may volatilize and be lost to the atmosphere as volatile gases. Ester form of 2, 4-D compounds are volatile and the gas injures susceptible crops such as cotton and tomato. In general, herbicide volatilizes and lost more rapidly from moist soil than from dry soil. Chlorpropham, diallate, EPTC and trifluralin are commonly lost from moist soil by volatilization.

Leaching

Leaching results in physical movement of herbicide molecules in soil. Excessive leaching leads to poor weed control due to removal of herbicide from their action site. Herbicides that are water soluble and remain dissolved in soil solution are readily subjected to leaching.

Surface Runoff

Heavy rain immediately after surface application of herbicides may often move the toxicant in solution from in surface runoff, away from target area. Water solubility and adsorption characteristics of herbicides inûuence extent of applied herbicide loss. Loss of herbicide by surface runoff is usually high in early stages of storm and from bare soils.

Majority of herbicides have relatively short life in the soil. Under warm and moist conditions, 2, 4-D persists for only 2 to 3 weeks and MCPA for 6-5weeks. However, 2, 4, 5-T may persist in the soil for 2 to 12 months.Picloram is one of the most persistent herbicides with its half-life varying from a few months in warm moist soils to four years in cold semiarid regions. Relative persistence of common herbicides in soil when applied at recommended rates is given in Table 2.8.1.

Table 2.8.1: Relative persistence of some common herbicides in soil when applied at recommended rates for weed control.

Less than three months	3-6 months	More than 6 months
Amiriotriazole Chlorpropham	Chlorbromuron	Atrazine
Cunazine	Diallate	Bromacil
Dalapon	Dinitroamine	Dichlobenil
Metoxuron	Isoproturon	Diuron
Prometryn	Linuron	Methazone
Propachlor	Methabenzthiazuran	Metribuzin
Propham	Metabromuron	Simazine
Terbutryn	Propyzamide	Terbacil
	Pyrazon	Trifluralin

Management of Herbicide Residues in Soil

An ideal soil applied herbicide should persist long enough to given an acceptable period of weed control but not so long that soil residues after crop harvest limit the nature of subsequent crops. Various management techniques have been developed to minimize residue hazards in soil.

Use of Optimum Dose of Herbicides

Application of herbicides at the lowest dosage by which desired weed control is achieved leads to minimum residue hazards. Application of fluometuron to cotton at recommended rates will not affect germination and growth of succeeding sorghum, finger millet, cowpea and lablab. Trifluralin at 0.72 and 0.96 kg/ha to soybean leaves no adverse residual effect on growth and development of succeeding wheat, but at 1.2 kg/ha, the wheat crop will be affected.

Cultural Practices: Ploughing with disc plough or inter-cultivations reduce the herbicide toxicity, as the applied herbicide is mixed with a large volume of soil and get diluted. In the case of deep ploughing, herbicide is inverted and buried in deeper layer to avoid its toxicity. Residual toxicity of propyamide, trifluralin and atrazine can be considerably brought down by ploughing.

Crop Rotation: Residues of herbicides applied to preceding crop should not be toxic to the succeeding crops in rotation. Application of atrazine tomaize, diuron to cotton and chlorben to soybean do not produce direct or residual effect on succeeding crops in cotton-maize-soybean rotation. Application of diuron at 1.0 kg/ha to cotton will not injure oats and wheat grown in rotation.

Use of Non Phytotoxic Oil and Activated Carbon

Atrazine residual hazard can be reduced by mixing it with non-phytotoxic oil. Activated carbon has high absorptive capacity due to its tremendous surface area.

Use of Herbicide Safeners

Safeners (antidotes) can protect the crop from possible damage by a herbicide. R 27788 marketed as a mixture with EPTC and butylate for soil application and napthalic anhydride (NA) is used for seed dressing on rice to protect the crop against molinate and alachlor. Another safener cyometrinil is used along with metolachlor in grain sorghum.

Leaching of the Herbicide

Leaching of the herbicide into the soil by frequent irrigations, is possible, especially, in the case of water soluble herbicides. Care should be taken toleach the herbicide beyond the root zone depth of the crop. Leaching may not be effective unless adequate drought drainage is provided to drain the water out of the field.

Effect of Herbicides on Environment

Herbicides enter the environment in the form of vapours and drifts from the application to the soil and plants. Herbicides in soil are subjected to various processes. They are liable to leaching to lower layers of soil or surface runoff with irrigation or rain water to storms, lakes, etc causing pollution problems. Use of herbicides for weed control in rice indicated no problem of herbicides residue in water samples collected from ûeld and drainage canals. Herbicides may affect soil microbial population for a short period, but eventually their population increase in few daystime after application.

Herbicide drift may be eliminated largely by using proper nozzle, spraying technique, proper time of application and use of correct formulation. Volatility is the characteristic of only certain herbicides such as anilines,carbamates, etc., which can be minimized to a great extent by proper soil incorporation and management practices.

Residues of Herbicides and Environmental Pollution

The length of time that a herbicide remains active in soil is called soil persistence or residual life of herbicide. The continuous and indiscriminate use of herbicide in various crops may lead to the accumulation of toxic residues in soil, underground water, crop produce as well as in environment. To a certain degree, persistence is a desirable property. Herbicides are claimed to persist in the soil as long as their biocidalactivity against weeds is wanted. Persistence exceeding this degree is called undue persistence.

Herbicides may be lost from soil either by physical removal of the unchanged molecule or by degradation. Their availability and thus activity, to plants and other organismsis related to their concentration in the aqueous and/or vapour phases. Various processes contributing to the persistence of herbicides in soil are demonstrated in Fig. 2.8.1 (Scheunert *et al.*, 1993).

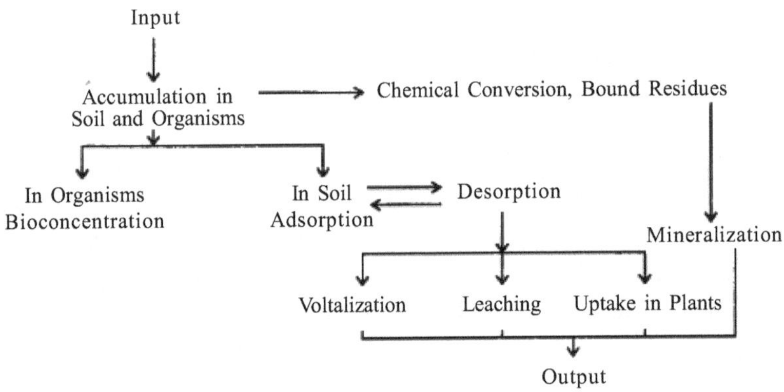

Fig. 2.8.1: Processes contributing to the persistence of herbicides in soil.

Degradation of Herbicide

Degradation is, generally, the major route for herbicide dissipation from soil. Herbicide is, gradually, transformed by degradation into series of intermediate metabolites, until the molecule is completely broken down. The intermediate metabolite is particularly important because they can have biological activity, toxicity characteristics and persistence which are different from those of the initial compounds. Degradation is possible in three ways: photochemically, chemically and enzymatically (microbically).

Photo decomposition is caused primarily by radiation at wave lengths between 200-400nm and affects the herbicide molecules characterised by weak bonds (diquat, praquat, trifluralin and chloroxuron) especially when the herbicide remain for a long time on the soil surface, particularly in case of heavy sunlight and high temperature conditions.

In general photochemical decomposition is considered to be insignificant for most soil residues in the field, particularly when herbicides have been incorporated by cultivation or leaching, as it is shown, for example, for some of the most recently developed chemical groups of molecules (sulphonylureas and imidazolinones) (Mengels, 1991).

Chemical degradation of the herbicides is caused by hydrolysis and oxidation reactions, sometimes followed by conjugation phenomenon. The degradation pathway depends both on the characteristics of the molecule (presence of certain chemical grouping) and on the soil and climatic conditions.

In particular, the colloidal content of the soil is very important because certain degradation reactions can be catalysed by mineral surfaces or functional groups in soil organic matter. Besides that, soil pH is of great importance because it influence molecular polarity and structure. Temperature and humidity also affect degradation. Also microbial degradation is carried out mainly by bacteria, actinomycetes and fungi and is one of the fundamental mechanisms of herbicide degradation.

Degradation is the major route for herbicide disappearance from soil. Photochemical degradation can be important for herbicides on plant surfaces but it is probably insignificant for most soil residues, particularly when they have been incorporated into the soil by soil cultivation or leaching. Under hot and humid conditions the herbicides may remain active in the soil over the short period of time. Persistence of isoproturon in soil varied widely. The half life of isoproturon were 43, 25 and 11 days in air dry, moist (field capacity) and flooded soils respectively (Kulshretha, 1982). In rice, degradation in relation to moisture status revealed that the dissipation rate of butachlor was faster at higher moisture level compared to low moisture level. In case of thiobencarb, it degraded at faster rate in the upland conditions compared to low land moisture regimes (TNAU).

Transformation Reactions in Soil

In general, in natural soils biotic processes prevail. In anycase, if the transformation products formed are persistent, they must be regarded as undesirable new xenobiotics in the soil.

Abiotic transformation reactions in soil comprise redox reactions, reactions catalyzedbyclay surfaces, and photolysis (Scheunen, 1992) and are given in Table 2.8.2.

Photolytic reactions are relevant for the soil environment only at the soil surface or inwater containing suspended soil particles.

Biotic reactions occur either within living organisms or are catalysed by extracellular enzymes present in soil. Processes may be oxidative, reductive, hyrolytic, or synthesizing new substances by reaction with natural compounds in the soil (Scheunert, 1992) which are presented in Table 2.8.3. Dechlorination reactions may be attributed to each of these mechanisms.

Table 2.8.2: Abiotic reactions of herbicides in soil

Type of Transformation	Reaction	Examples of herbicide for primary metabolites
Redox reactions	Oxidation	Anilines
	Oxidative coupling Reduction	Phenols, Anilines
	Mineralization	Trifluralin, Pendimethalin
	Hydrolysis	Amitrole
Reactions catalyzed by clay surfaces	Ester formation	S-Triazines
	Oligomerization/	Alcohols, Alkenes
	Polymerization	Olefines, Dienes
	Direct photolysis	
Photolysis	Reaction with reactive oxygen species	Fluchloralin, Bentazone Olefines

Table 2.8.3: Biotic reactions of herbicides in soil

Type of Transformation	Reaction	Examples of herbicides (or primary metabolites)
Oxidative processes	C-hydroxylation	Aromatic compounds
	B-oxidation	Phenoxyalkanoic acids
	C-cleavage	Aromatic compounds
	N-oxidation	Anilines
	N-demethylation	Phenylureas
	Ether cleavage	Phenoxyalkanoic acids
	Mineralization	Nearly all organic compounds
Reductive processes	C-reduction	Alkenes and alkines
	N-reduction	Nitro compounds
Hydrolytic processes	Ester hydrolysis	Carboxylic esters, sulfates
	Nitrile hydrolysis	2,6-Dichlorobenzonitrile
	Epoxide hydrolysis	Intermediates of aromatic C-hydroxylation
Synthetic processes	Conjugation	Butachlor, Dichlofop-phenol
	Reaction with natural humic monomers	Chlorinated anilines and phenols
	Interaction between different pesticides	Chlorinated anilines and phenols

Mineralization

Mineralization of herbicides in soil, resulting in carbon dioxide and other small inorganic fragments, is a complex reaction. It may be achieved by metabolism of microorganisms growing on the respective herbicide as a sole carbon source; in other cases, the first steps towards mineralization are performed by one microorganism and the following by others; abiotic processes may be involved also. Table 2.8.4 gives some examples of $^{14}CO_2$ evolution from the ^{14}C-labelled pesticides in a soil-plant system during one week. It shows that mineralization of many of them is very slow. Pentachlorophenol is mineralized relatively rapidly despite of its chlorine atoms. This is due to its aromatic hydroxyl group which promotes enzymatic attack.

Effect of Herbicide Residues on Succeeding Crops (Soil Persistence)

The persistence of herbicides in the soil strongly inûuence the choice of the succeeding crop. This problem arises with rotation crops and, above all, with replacement crops, which have to replace a crop lost, due to adverse weather conditions, parasites or disease. Excessive herbicide persistence increases the risks of carry-over for the successive crop. As residues tend to be near the surface, the risk of damage is greatest when seeds are sow without ploughing into the herbicide layer. Ploughing greatly reduces the risk of damage because it removes most of the herbicide from the vicinity of the seed and young roots and also dilutes the residual concentration of herbicide in the plough layer.

Data presented in Table 2.8.4 gives a simple guide to the maximum persistence of actual residues of herbicides applied at normal doses and times (Hager and Nordby, 2007), An ideal herbicide is that which persists in soil for a period of 6 to 8 weeks approximately or it must degrade to undetectable limits before the harvest of the crop to which it was applied. So soil persistence of herbicides depends upon type of herbicides, dose, soil type, organic matter content, rainfall, temperature, light, soil microflora, soil reaction etc. The recommendation of herbicides should be according to crop rotation in order to avoid its adverse effect on the succeeding crop(s).

Table 2.8.4: Soil persistence of some common herbicides

Duration	Herbicides
< I month	2, 4-D, MCPA
1 to 3 months	Acetochlor, Alachlor, Ametryn, Bentazon, Butylate, DCPA, Dirnethenamid, EPTC, Flumetsulam, Foramsulfuron, Halosulfuron, Lactofen, Linuron, Mesotrione, Metolachlor
3 to 12 months	Atrazine, Benefin, Bensuiide, Bromoxynil, Chlorimuron, Clomazom Diuron, Ethalfluralin, Fomesafen, Hexazirione
>12 months	Bromacil, Chlorsulfuron, Imazypyr, Picloram, Prometon, Sulfometuron, Tebuthiuron

Residue of Herbicides in Crop Produce

Field studies were conducted on loamy sand soil at Ludhiana during 1989 to study persistence of isoproturon (0.5 and 0.75 kg/ha) and methabenzthiazuron (0.875 and 1.31kg/ha) in onion bulbs by Randhawa and Sandhu (1992). They reported faster degradation of both these herbicides in onion as compared to soil. Both herbicides degraded to non-detectable limit in onion bulbs within 30 days.

Gas chromatographic analysis of butachlor residues (Jayakumar and Sankaran, 1988) applied to paddy field showed that at lower rate (1.0 kg/ha), butachlor persisted upto day 30 while at higher rate (4.0 kg/ha) the herbicide persisted up to harvest.

The contents of herbicides in edible parts of the crop must be below the maximum residue limits (MRL) and these values for import crops are given in Table 2.8.5 (Hager and Nordby, 2007).

Table 2.8.5: Maximum residue limits (MRL) permitted in various crops

Herbicides	Crop	Max. Residue Limits (ppm)
Triazine	Soybean	0.15 to 0.25
	Alfalfa	0.04 to 0.10
	Oats	0.06 to 0.15
	Wheat	0.075 to 0.18
Dinitroanaline	Corn	0.1 to 0.2
	Wheat	0.2 to 0.3
Clomazone	Corn	0.05 to 0.2
	Wheat, Alfalfa	0.015 to 0.1
Imazaquin	Corn	0.002 to 0.01
	Wheat	0.01 to 0.03
Imazethapyr	Corn	0.01 to 0.03
	Sorghum	0.004 to 0.015
Chlorimuron	Corn	0.001 to 0.002
	Wheat	0.002 to 0.005

Estimation of Herbicide Residues

Residues of applied herbicides in soil, crop produce, plant parts, underground water etc. can be estimated with the adoption of the following techniques.

Determination of Herbicide Residue by Volumetric Method

In this method the substance is allowed is to react with a reagent of known strength and from the volume of the reagent used for the complete reaction of substance, the amount of the substance present is determined.

Methods

- The preparation of one or more standard solutions of accurately known concentration

- The use of accurately graduated vessels by which volumes can be measured
- Some means of recognising the end point, i.e. the completion of reaction volumetric analysis involves titration. Titration is a process whereby a standard solution is brought into reaction with another solution. The point at which reaction is exactly completed is called end point or equivalent point. The titrimetry is based on the fact that the product of volume (v1) x concentration (n1) of reaction substance is equal to the product of volume (v2) x concentration (n2) of another reacting substance involved in the titration. That is $V_1N_1 = V_2 N_2$

Methods of Titration

1. **Direct method** : Substance is directly determined by titration with a standard solution upto the end point
2. **Indirect method**: In this an excess quantity of the substance over and above that required for the reaction is added and the concerned excess quantity is determined by back titration.

Testing for Herbicide Residues

If herbicide carryover is suspected, a soil chemical test or a bioassay can be used to determine if harmful levels of herbicide are present. Chemical analysis can be expensive, so a bioassay conducted either in the suspect field or in a warm, sunny indoor location (such as a greenhouse) may be more feasible. These tests help predict potential herbicide-residue problems so the grower can make better decisions about crop rotation, herbicide selection, planting date, and other cultural practices.

Bioassay

Bioassay is a measurement of biological response of living organism to determine thepresence and concentration of chemical in the soil. In spite of rapid developments in analytical method, bioassay remains a major tool for qualitative and quantitative determination of herbicide residues.

The use of bioassays to determine the concentration of an herbicide in soil or other growing media has received much attention and has been of immense importance to research in weed science. Although chemical methods of detecting herbicides can often determine smaller amounts with greater precision, bioassays offer several advantages (Rahman *et al.*, 1993).

Bioassay is a simple, inexpensive, accurate and direct method of determining if it is safe to grow crops on farmland previously treated with known herbicides or on crop land with an unknown history of herbicide use. A bioassay can detect if herbicide or chemical residues are present in the soil at concentrations high enough to adversely affect crop growth, yield and quality.

Bioassays Should be Conducted

- When newly seeded or established plants exhibit abnormal growth injury
- When seeding or planting on areas previously treated with herbicides known to be residual, such as imazethapyr.
- When using abandoned farmland, which may contain herbicide residues such as picloramor triazines.
- When using non-cropland, such as railways, roadsides, storage areas, and industrial sites that may contain residues of chlorsulfuronand picloram.
- When purchasing livestock manure or topsoil of unknown origin.

In all cases of herbicide use, refer to the manufacturers guidelines to determine if herbicide residue is a concern.Figure2.8.2 shows a typical bioassay being performed on several species to detect bromacil residues in soil.

Fig. 2.8.2: Bio assay in bromacil residue soil

Collecting and Submitting Soil Samples

The accuracy of a bioassay depends on two factors:

- Sampling technique
- Depth and area of soil sampled

For best results, soil samples should be collected and the test conducted about one month before seeding time.

To collect soil samples

- Sample soil from areas suspected of having herbicide residues as well as those areas believed to be free of herbicide. Keep samples separate.

- Take separate samples from high spots and low spots, including areas where sprayer overlap could have resulted in an overdose, particularly if the overlap is till visible in standing stubble.
- Collect soil samples from the upper surface (0 to 8 cm/0 to 3 inches). Most residual herbicides are bound in the upper top of soil (5 cm/2 inches), unless the land has been cultivated to a specific depth, in which case, samples should be taken at the depth of cultivation, usually 10 to 15 cm/4 to 6 inches.
- Using a spade, trowel or soil sampler, take several samples from the suspect area and combine them. Collect enough soil, about 2 kg or 4 lbs, to fill several 10 cm (4 inch) pots.
- Along with the soil sample from the herbicide-free area, you may now take one to three additional soil samples, depending on land topography.
- Store the samples in cool conditions. If the soil is wet, spread it out to dry, keeping different samples separate to avoid cross contamination. When dry, crush clods to pea-sized particles.

Table 2.8.6: Selecting plant species for herbicide bioassay

Herbicide group	Test species	Common InjurySymptoms
Group 2		
imazethapyr *chlorsulfuron*	Sugar Beet, Canola	Stunting, purpling, general yellowing of new growth, terminal bud death
Group 3		
trifluralin,ethalfluralin	Green Foxtail, Oat	Stunting, swollen andshortened roots
Group 4		
picloram,clopyralid	Faba Bean, Flax	Malformation; bendingand twisting of stem, leafcupping and curling
Group 5		
triazines, *bromacil*	Oat, Cucumber	Stunting, interveinalyellowing of new leaf
Group 8		
triallate	Tame Oat, Wild Oat	Lack of emergence, brittleand dark greencoleoptiles, shoots mayform loops

I. Seedling Transplant Bioassay Method

A simple pot method has been used successfully for determining the availability of §'\'€I'8.I photosynthetic inhibitor type herbicides in soils (Stalder and Pestemer, 1980). With same modifications, this method has been further developed for measuring the biological activity and plant availability of sulphonylurea herbicides.

Turnip (*Brassica rapa*sp. *Rapa*) was found the most suitable test species for the sulphonylurea herbicides for this method. Seeds were sown uniformly in vermiculite in seed rays. After 7-8 days growth, seedlings can be transplanted into pots for conducting residual studies. This method is very quick and simple.

Seedling plant bioassay technique is also helpful in resistance studies. In case of any doubt regarding resistance of a weed species to a particular herbicide, seedlings of that weed can be transplanted in pots kept in labs or green house and herbicide(s) can be applied after the establishment of seedlings. On the basis of results, farmers can be guided for proper use of herbicide.

II. Direct Seeding Bioassay Method

Seeds of indicator plants (Table 2.8.7) are selected according to the nature of study. Counted number of seeds is sown in each treatment. Mortality percentage and dry mater accumulation are two major parameters which decide the residual concentration of any herbicide in the soil.

Selection of Test Plants (Bioassay material)

Ideal test plants must be fairly sensitive to detect very low herbicide concentration. Also they should show a gradual increase in susceptibility under increasing herbicide concentration and depict sensitivity on easily observable parameters.

Table 2.8.7: Bioassay materials used for various groups of herbicides

Herbicide group	Bioassay species
Aliphatic halogenated acids e.g. dalapon	Oats, millets, cucumber, barley, wheat, rice
Acetamides e.g. diphenamide	Crabgrass, oats, barley, ryegrass, pig weed
Acetanilide- alachlor, metolachlor	Cucumber, ryegrass, crabgrass
Benzoic acid	Cucumber, oats, foxtail, pig weed.
Dicamba	Beans, sorghum, cucumber
Chiorpropham	Cucumber, oats, ryegrass
Phenoxy derivative e. g. 2,4-D	Cotton, pig weed, tomatoes, mustard
Substituted ureas-diuron, isoproturon	Cucumber, ryegrass, oats, barley, millet, sorghum
Thiocarbamates e. g. EPTC	Ryegrass, oats
Dinitroanilines e.g. trifluralinfluchloralin	Oats, sorghum, rice, cucumber
Triazines e.g. atrazine, simazine	Oats, cucumber, sugar beets
Sulphonylurea	Turnips, sugarbeet, bajra, sorghum

The response of indicator plants to herbicides can be evaluated by the following ways:

Germination test: Many herbicides strongly inhibit the germination of sensitive species.

Plant part response: Which includes measurement of reduction/elongation in shoot length, fresh and dry weight of root and shoot chlorosis or necrosis and morphological abnormalities.

Whole plant response: Which includes measurement of general chlorosis, morphological abnormalities determined by fresh and dry weight of whole plant. Generally the tops (shoot portion) only are weighed as separation of roots from soil is laborious work. Leaf area of test plant can also be measured. Bioassay experiment should only be terminated when progress in toxicity symptoms if any has stopped completely.

Methodology in Conducting Bioassay Using Higher Plants

Standard curve

Mix thoroughly the standard amounts of herbicides (0, 0.1, 0.2, 0.4, 0.8, 1.6 and 3.2 ppm) into the soil taken from control plot or area adjoining to the experiment. Dibble 10 seeds of test species in each pot at uniform depth (2 to 2.5 cm). Replicate the experiment to suit the statistical analysis. Take the germination count after 7 days and thin to 2 to 3 plants uniformly to avoid the competition effect. Observe for the malformations and epinasty symptoms in pulled out seedlings. Two weeks after sowing take the plant height and after 4 weeks cut the plants close to the soil surface and take the dry weight. Draw the standard curves for the above parameters using the concentrations of the herbicide on the X axis and the plant parameters on the Y axis.

Determination of herbicide residue in soil

Take the soil sample to be analysed from the field and mix thoroughly. Weigh 1kg soil into the plastic container and dibble 10 seeds of test species along with the standards. Record germination percentage, plant height and dry matter production under similar conditions as used in establishing the standard curve. Compare the values of each parameter from the standard curve and express the result.

Precautions

- Mix the collected soil samples thoroughly and impose the treatments.
- Water the pots uniformly without draining.
- Dibble the seeds of test species at uniform depth.

Assessment Parameters

The response of indicator plants to herbicides can be evaluated in various ways.

Germination tests

Many herbicides strongly inhibit the germination of sensitive species but few tests have been based simply on germination numbers. Sub lethal concentrations frequently produce an inhibition of radical or shoot elongation, which is dose-related in a range sufficiently large to allow valid measurements. Typically, the root or shoot elongation is observed after a period of 24 to 96 hours.

Assessment of Plants

Determination of dry weight is most common assessment used in bioassay. Generally, the tops only are weighed since the separation of roots from soil is laborious. Observations based on fresh weight and dry weight generally give good estimations.

For many photosynthesis inhibitors shoot growth is reduced before the appearance of the injury symptoms. Measurement of plant height or leaf length provides an assessment of herbicidal activity, which may replace weighing. While leaf length is easily measured on monocots, other measurements may be considered on other species. For instance, the petiole length of the first trifoliate leaf of legumes gives dependable estimation. Observations of plant height generally give similar results. However, with herbicides inducing deformations in the shoot, the plant height values should be used with care. Koren *et al.* (1968) found that with thiocarbamates, the height reduction was greater than weight reduction. Recording both the parameters is useful in experiments including different type of herbicides (Horowitz and Blumenfeld, 1973).

Physiological and Morphological Effects

The physiological, biochemical and structural modifications of plants induced by herbicides can be used for bioassay assessment. Reduction of photosynthetic activity by photosynthesis inhibitors has been measured on leaf discs (De Silva *et al.,* 1976). Chlorosis may be evaluated by determining chlorophyll content through spectrophotometer (Horowitz, 1970).

Symptoms

Symptoms, which are typical of a certain group of herbicides or of a given compound, can be used for qualitative assay and if the intensity of symptoms is dose related, it can also be used for quantitative determinations. For example, epinasty and stem deformations of tomato and malformations of leaves of cotton have been used to measure the effect of 2, 4-D.

Testing for Specific Herbicide Groupsusing Bioassay Technique

Triazine Residues For suspected carryover from triazine herbicides, such as atrazine and Princep (simazine), an oat plant bioassay works best. Place about 15 oat seeds in each container of soil and cover the seeds with about 1 inch of soil. Wet the soil with water, but do not saturate it. Place the containers in a warm location (70° to 75°F) where they can receive ample light. Sunlight is essential for the development of the plant, as well as for inducing symptoms of triazine injury. The container should be watered as needed. Injury symptoms should become apparent within 10 to 14 days after emergence.Triazine injury is characterized by chlorosis (yellowing), then necrosis (browning) of leaf tissue. As injury symptoms start at the leaf tip and develop toward the base, a comparison with the plants in the check soil is essential. If injury appears on the oats, enough herbicide residue may be present to injure a susceptible crop. Planting a more tolerant crop is suggested. In general, the order of susceptibility from most to least susceptible to triazine herbicides is as follows: Ryegrass > Alfalfa > Oats > Wheat > Soybean > Sorghum > Corn

Laboratory Analysis

Laboratory analysis involves extracting herbicide from the soil with the use of specialized equipment to detect very small amounts. The amount is usually expressed in parts of herbicide per million parts of soil (ppm). This measurement can be transposed into pounds of herbicide active ingredient per acre (lba.i./ A) if we assume that an acre of soil weighs 1 million pounds in the top 3 inches and 2 million pounds in the top 6 inches. For a soil sample taken to a 3-inch depth, 1 ppm = 1 lb/A of residue. For a soil sample taken to a 6-inch depth, 1 ppm = 2 lb/ A of residue. A lab report of 0.2 ppm atrazine, then, means that there is 0.2 pound of atrazine per acre if the samples were taken to a 3-inch depth, and 0.4 pound per acre if taken to a 6-inch depth. The location and concentration of the chemical depend on the herbicide used, the soil type, whether the ground was tilled, and the amount of rainfall since application. In most medium-textured soils (silt loams, silty clay loams, sandy clay loams), the herbicide re- mains primarily in the top 3 inches unless there was excessive rainfall, the ground was ploughed, or the herbicide was deeply incorporated. If the soil has a high sand content (coarse texture), then herbicide leaching may be greater. Movement of the herbicide from the surface soil zone by tillage or by rainfall decreases the likelihood of crop injury. The risk of injury is greater when the herbicide residue is concentrated in the top 3 inches rather than distributed throughout a 6-inch soil depth. Therefore, it is best to sample in two sepa- rate sections, from 0 to 3 inches and from 3 to 6 inches. Whether parts per million or pounds of active ingredient of herbicide per acre is used, it is difficult to translate these units of measure into potential crop injury. Many variables affect crop susceptibility or tolerance, including soil type, crop sensitivity, and environmental conditions after planting. Crop injury is more likely on more coarsely textured soils or under cool, wet weather conditions. Although you are cautioned that crop injury may still occur below these levels. Laboratories may differ in available tests and in the prices for analysis. The cost can range from $20 to $200 per sample for herbicide analysis. Most laboratories can analyze a sample and have the results in 5 to 7 days. Contact your local Extension office for more information on laboratory selection.

Correcting for Herbicide Residues

If the lab test or bioassay indicates a potential herbicide residue problem, several steps can be taken. 1. First select a tolerant crop or variety. This selection depends on what herbicide is of concern. Check current herbicide labels for more information on crop tolerance. 2. Tillage can help dilute herbicide in a problem field. 3. Plant the field that concerns you last. Delaying planting allows more time for the herbicide to dissipate. 4. If the triazine herbicides or chlorimuron is sus- pect, be sure to check the soil pH and adjust your management practices accordingly. 5. If imazaquin or imazethapyr is suspect, check for low soil pH (< 5.5). Liming would both benefit crop growth and minimize carryover of these herbicides. Soil bioassays or laboratory tests are not 100 percent accurate in predicting herbicide-residue problems. Crop response to herbicide residue depends on various factors, including species and variety, soil type, and environmental conditions after planting. So, predicting crop injury is often difficult. However, using a soil chemical test or bioassay can help in deciding whether a potential problem exists and in choosing the appropriate crop or variety.

Advantages

1. Handy – tedious operations are not needed
2. More rapid
3. Fairly accurate
4. Economical ,involves lesser cost in chemical and apparatus
5. Minute quantities can be worked easily which would be otherwise difficult to weigh
6. It is possible to estimate one or more substance in the presence of others without effecting separation.

Chemical Analytical Methods for Determination of Herbicide Residue

Herbicide residue Determination by Spectro-photometric and Chromato Graphic Methods

Monitoring herbicide residues in environment and crop produce are important in the standpoint of healthy environment. Herbicide residue analysis is a special task involving different steps such as sampling, extraction, clean up and determination. The various steps involved in the residue analysis are discussed below.

Methodology

The herbicide residue analysis involves the following steps viz., sampling, extraction, clean up, concentration and determination.

Sampling

The size of the sample to be taken for residue analysis is very important. One must take the sample large enough that the residues in the final samples for determination should be above the minimum detectable level i.e. if instruments can measure a quantity of 0.1 $\mu g\ g^{-1}$ than final sample for determination should contain more than 0.1 $\mu g\ g^{-1}$. This sample quantity can be determined logically based on the dose and time of sampling. If the residue analysis is taken up immediately after application, 25-50 g of sample is sufficient. In the case of harvest stage samples may be from 500 g to 1.0 kg or even more. In the case of other samples such as cloths, the sample size may be very high. Contamination in any way in the samples must be avoided following good laboratory practices.

Extraction

Almost all the herbicides are organic excepting a few inorganic compounds. The organic compounds are usually extracted using organic solvents. Based on the polarity of the organic compound (herbicide) the solvent is chosen. Either single or two or three solvents are used for extraction depending upon the type of the compound. Polar compounds are extracted with polar solvents like acetone, acetonitrile, methanol etc., and non-polar compounds are extracted with non-polar solvents such as hexane, chloroform, dichloromethane etc. The critical idea in extraction is that whatever solvent or number of solvents used, the organic compound to be analysed in question is to be fully extracted.

The quantity of the extractant used for extraction is up to the substance submergence level or more. Normally 2 minutes blending in a blender or 30 minutes shaking in reciprocating shaker or keeping over night is equal. Depending upon the sample, one or the other or all of the above extraction methods may be followed so that the extraction is to the fullest extent. The extracted material is filtered using Buckner funnel under vacuum or using normal filtration techniques or centrifuged at 2000 rpm for 2 minutes to remove the particulate matter. The filtrate is further subjected to clean up process. Often the filtrate is concentrated to a lower volume using rotary vacuum flask evaporator before cleanup process. Normally two to three extractions are done for better extraction of the compounds.

Cleanup

This step is required in the case of herbicide residue analysis because while you are extracting the organic compound (herbicide) not only the compound of your interest will be extracted but also the other undesirable organic compounds which are soluble in extracting solvents are extracted, called as co-extractives, which normally will interfere during the process of determination. The cleanup may be accomplished by the following methods.

Partitioning

Based on the polarity of the compounds and co-extractives (impurities) these techniques are being adopted. For example, if a compound is polar and impurities are non-polar then portioning the extracted compound using non-polar impurities and vice-versa. To increase the polarity, normally water is added or sodium chloride at different concentration or even up to saturated levels are being used.

Column Cleanup

Glass columns of 18 mm of any length of above 30 cm are used for this purpose. The column is packed with absorbing material such as alumina, silica, florisil etc., along with activated charcoal (decolouring agent) and anhydrous sodium sulphate (dehydrating agent). In the packed column the concentrated extract is poured and eluted with suitable solvents. The molecular configuration and weight of the compound play an important role here. The low molecular weight compounds will be eluted first and based on this technique the compound is cleaned up.

TLC Cleanup

Thin layer chromatography is an excellent tool for cleanup of the samples. Over the silica gel coated plates the concentrated solvents are spotted along the standards. The eluted compound is identified using standard and is separated by scrapping and dissolving in suitable solvents.

Concentration of the Extract

The concentration of the extract is usually done with the help of a rotary vacuum flask evaporator where in at low pressures the solvent evaporates faster.

Determination

Since a wide range of chemicals is used as herbicides, a variety of methods are used in their analysis.The instrumentation methods involve visible and ultraviolet spectrophotometry, thin layer chromatography, gas chromatograph, high performance liquid chromatography etc. The visible and UV spectrophotometry are useful when the herbicides are not amenable to chromatographic analysis. Thin layer chromatography is a quick, cheap and efficient technique for qualitative and quantitative analysis and for cleans up. Gas chromatography is undoubtedly the most versatile and sensitive method. HPLC is finding applications in herbicide analysis especially when other instrumentation techniques are not useful besides there is no limitations set by volatility, a wide range of compounds can be separated and analysed.

UV Spectrophotometry

The sample to be analysed must be free from any material that would absorb light at or near the same wavelength as the chemical under study. This is essential for any accurate spectrophotometry method. Ultraviolet spectrophotometry measurements offer several real advantages to the residue chemist. The radiation absorbed is characteristic of the material, thereby offering some specificity to the measurement and sometimes clues to the identities of unknown compounds. Within the limits of Beer's law, the degree of absorption of radiation is proportional to the concentration. Many herbicides have large absorptivities and these can be analysed at very low concentrations. Because many spectrophotometric methods are nondestructive, the absorbing material can often be retrieved for further scrutiny by other types of instrumentation.

Herbicide residue analysis requires an effective cleanup process prior to the actual determination. This cleanup is the most difficult aspect of the entire procedure. Ultra violet methods require special vigorous cleanup to ensure that final solution is free from any other material that will absorb light in the region of the spectrum where the herbicide content is to be measured.

Procedure

Spectrophotometric method employing the ultraviolet and visible region have been used in quantization for pesticide. The 185-400 mu region is the one used for UV analysis while 400-800 mu for analysis in the visible region. Visible range measurements include the treatment of a substance (pesticide) in solution with a reagent that causes a development of colour and absorbance of which is related to the concentration of the substance being determined. The relationship between absorbance and concentration of substance is normally linear. The fundamental relationship between the light absorption of solution is given by Beer

Lambert law.

$I_t = I_0 e^{-kcd}$ or $\log (I_0/I_t) = kcd$

I_t = Intensity of transmitted light; I_o = Intensity of incident light;

d = thickness of absorbing medium I_o/I_t as transmittance

Kc = constant T; reciprocal I_o/I_t as absorbance A (A = I/T)

Thin Layer Chromatography

Thin layer chromatography is a quick cheap and efûcient technique for qualitative andquantitative analysis and for clean up method prior to GLC or HPLC.

TLC is used mainly for confirmation of residues following initial screening and quantification. Confirmation by TLC, which is based on comparison of migration distances of the herbicide of interest with authentic standards run on the same layer. In addition, the quantification can be carried out by TLC if a gas chromatograph is not available or if the herbicide of interest is unstable during determination. Extraction, cleanup and concentration steps normally precede TLC determination. The advantages of TLC are simplicity, rapidity and low cost. Sensitivity ranges from about 5-500 ng for most herbicides detection methods. Rapid semi-quantitative estimation can be achieved by visual comparison of sample and standard spot sizes and/or intensities, and more accurate quantification can be carried out by *insitu* scanning of spots with a spectro densito meter. Spots are applied to the thin layers using simple dispersible capillaries, GC syringes or automatic multiple spotting devices, standard solutions must be spotted on the same plate as the sample, preferably on both sides of the sample spot.

Layers are hand coated with a commercial adjustable spreading device or layers pre-coated on glass, plastic or aluminium backing may be used. Analytical layers are usually 250µ thick. Pre-coated layers are of high purity and uniformity and are used almost exclusively in most laboratories, especially for *insitu*quantification by densitometry. Activation of adsorbent layers (e.g. 80-110°C for 30-60 minutes) prior to spotting and development is often required. Once the adsorbent has been activated, it must either be used promptly or stored under desiccating conditions or activation must be repeated. Silica gel and alumina layers usually give the best results, but polyamide, microcrystalline cellulose, ketselguhr, zinc carbonate and magnesium oxide are among other adsorbents have also been used. Chromatography is carried out, in a development chamber, most often a rectangular glass paper-lined tank saturated with solvent vapors. Ascending development for a distance of 10-20 cm is typical. It is important to follow exactly all stated conditions when attempting to reproduce a separation. The temperature, development chamber design and equilibration and water content of the adsorbent are probably the most frequent sources of variation among laboratories.

The technique termed 'High Performance Thin Layer Chromatography' (HPTLC) has become increasingly important for separations and *insitu* quantitative analysis in the recent past. HPTLC is carried out on 10 x 10 cm or 5x5cm layers of silica gel with a smaller particle size and a narrower particle size distribution than in conventional TLC plates and thereby gives improved resolution and sensitivity of detection. Volumes not larger than 1 ìl must be spotted to realise best results. For manual application, spotting is usually done with a pt-Fe tipped nano pipette or equivalent or this type of pipette is used with an automatic spotting device that control both the pressure of the pipette tip on the layer and the duration of contact. Solvent development is carried out in the miniature glass rectangular chamber. High resolution is achieved rapidly with short development distances.

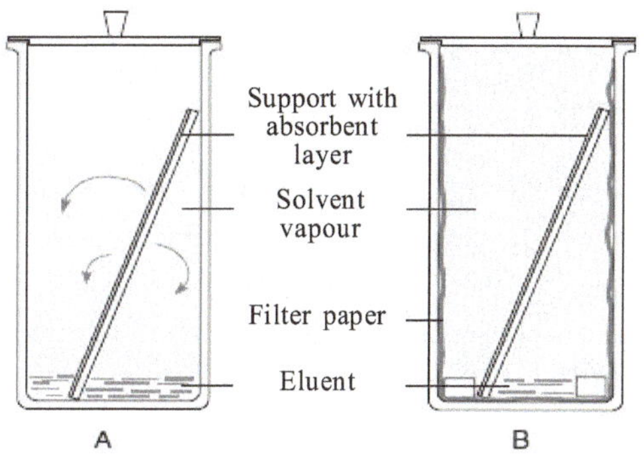

Developing chambers and saturation

A) Chamber with normal saturation. The arrows stand for the evaporation of the eluent from the layer and the dots symbolise the vapour density.

B) Chamber lined with filter paper, saturated with eluent vapour

TLC chamber

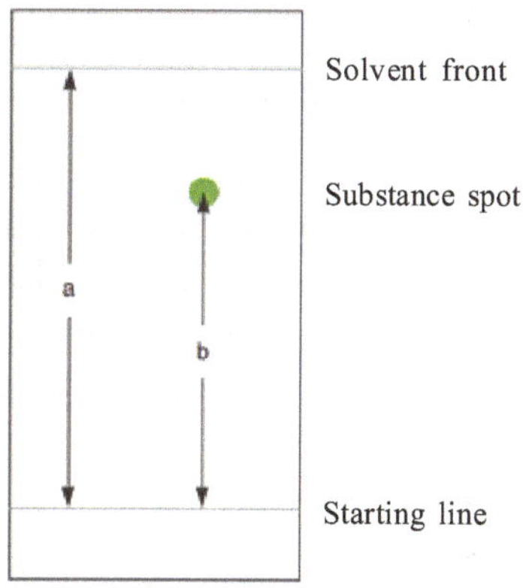

Movement of solvent & Substrate

Evaluation of a Thin Layer Chromatogram

The evaluation depends on the purpose of a chromatographic analysis. For qualitative determination often localization of substances is sufficient. This can be easily achieved by parallel runs with reference substances. A parameter often used for qualitative evaluation is the *R*f value (retention factor) or the 100fold value h*R*f. The *R*f value is defined as follows: i.e. the *R*f values are between 0 and 1, best between 0.1 and 0.8 (i.e. 10-80 for h*R*f). If reproducible *R*f values are to be obtained it is, however, essential that several parameters such as chamber saturation, constant composition of solvent mixtures, constant temperature etc. are strictly controlled. Qualitative evaluation is generally made directly on the TLC plate via the characteristic *R*f values of substances, i.e. the ratio of distance start-substance zone to distance start-solvent front and specific chemical reactions.

Disadvantages

The detection sensitivity of TLC, qualitatively as well as quantitatively, depends heavily on the pre-chromatographic or post-chromatographic sample treatment and with fluorescence measurement reaches at least the picogram level.

Gas Chromatography (GC)

In herbicide analysis, GC has assumed a role of prime importance as compared to other methods. This is because GC technique is capable of rapidly resolving complex herbicide mixtures and provides qualitative identification and precise quantitative analysis of the compounds. In multi-residue analysis, GC is the only method of choice. The GC unit consists basically of the following elements: carrier gas system, injection port, chromatographic column, oven, detector and recording system.

Carrier Gas System

The carrier gas ensures the migration of the components of the sample to be separated. The nature of the carrier gas in principle is any gas, which does not interact either with the stationary phase or the components of the sample. The gases that are widely used in gas chromatograph are nitrogen (this is easily available and handled), argon and helium.

Injection Port

The task of sample injector system is to introduce the gas, liquid or solid sample rapidly and in a reproducible manner into the column. The syringes manufactured by Hamilton Ltd., are the most widely used and they are marketed with 1 to 500 ìl volume. Syringes of 1, 2 and 5 ìl volumes are the most common in gas chromatography and with 1 ìl syringe a sample of 0.01 ìl can be directly measured.

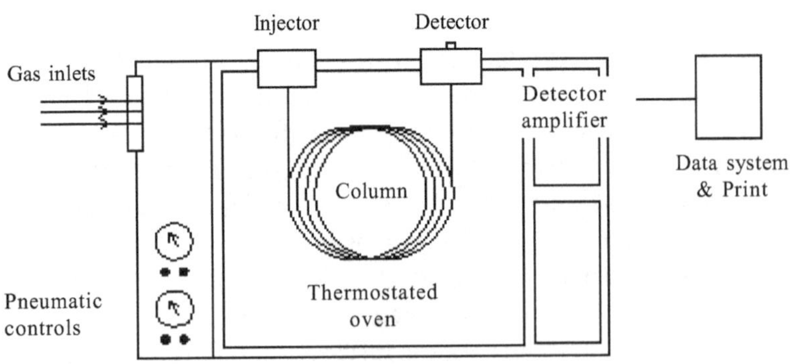

Schematic Diagram of Gas Chromatography

Chromatographic Column

With respect to their dimensions and arrangement of two basic types of columns, the packed and the open tubular (capillary) column are rather different, though the latest type (packed capillary columns) represents a certain transition between the two.

Table 2.8.8: Types of chromatographic columns

Column designation	Chemical name of stationary phase material	Equivalent designations
DC-200	Methyl silicone	OV-1,OV-I01,SE-30,SP-2100, DC-11, SF-96
QF- 1	Trifluoropropyl methyl silicone	OV-210, SP-2401
SE-52	Methyl silicone 10%phenyl substituted	OV-3
XE-60	Cyanoethyl methyl silicone	OV-225
OV-17	Methyl silicone 50% phenyl substituted	SP-2250
OV-7	Methyl silicone 20% phenyl substituted	
DEGS	Di ethylene glycol succinate	

Oven

The GC column is enclosed in an oven made of metal block. The temperatures of the injection block, column, oven and detector can be varied and controlled. Injection block temperature is generally 20 to 50 °C higher than the column temperature. In most commercial instruments, oven temperature can be controlled to at least+ 1 °C or better with a gradient of + 2 °C.

Detectors

Detectors have the task to sense continually, rapidly and with high sensitivity the components, which appear in the carrier gas as it emerges from the column. The detector senses the changes in certain physical or chemical property of the eluent gas stream on the appearance of the components. With the help of an appropriate current source and signal transmitted, the detector gives a direct electric signal or the response of the detector is converted into an electric signal by an appropriate electronic system. Usually this electrical signal is a voltage, which may be led directly or through an amplifier for recording or data processing.

Performance of Selective GC Detectors to Different Herbicides

Functional group detection limit and application for different detectors are furnished in Table 2.8.9.

Table 2.8.9: Performance of GC detectors

Detector	Element or functional group	Detection limit(pg/sec)	Herbicide analysis application
ECD	Trichloro nitro	0.50	2,4-D and 2.4.5-T esters, polynuclear aromatics, nitro herbicides
	Polyhalo-genated	0.05	OC compounds, pentachlorophenol, dioxins
FPD	P	0.20	OP compounds,
	S	4.00	Carbamates containing S, ethylene thiourea
AFID/ TID	P	0.02	OP compounds
	N	0.20	Carbamate insecticides, triazines, urea and nitro herbicides, nitrosamines
HECD	Cl	0.50	Phenoxy herbicides
Hall -EC	N	1.00	Carbamate insecticides, triazines, urea and nitro herbicides, nitrosamines

Recorders and Integrators

Recorders

The response of the detector is plotted by the recorder, which furnishes the chromatogram, the result of the analysis. In recorders, electronic chart potentiometers are commonly used. They operate on the principle of balancing the input signal by a feedbacksignal of the same intensity but of the opposite sign. The difference between the input and feedback signals reaches through an amplifier to a serve-motor, which moves the pen. The recording pen has a response speed of 0.25 to 2 sec for full-scale deflection. Accurate measurement of peak areas can be done by ball and disc integrators and electronic analogue integrators.

Integrators

Integrators are nowadays used increasingly to process GC data. Measurement of peak areas and retention time with high accuracy, correction for base line drift and analysis of partially resolved peaks by a curve fitting procedure to obtain percentage of each component are some of the operations that are carried out by computers.

Gas Liquid Chromatography (GLC)

GLC has been most widely accepted analytical technique in estimation of pesticide residues. The sensitivities of detectors is 0.1-1 ng of pesticide which is equivalent to 0.01-0.1 ppm residues in 10g samples. Chromatography is a tool to separate mixtures of solutes based on the differences in physical properties. The separation is brought about by their differential migration through a porous medium under the inûuence of a moving phase. In GLC the stationary phase is a liquid and mobile phase a gas. The GLC unit consists of the following components:

Carrier Gas: Commonly used gases are N2, He and H2. Nitrogen IIOLAR Grade I or I1 is recommended for ECD-GLC.

Columns: Borosilicate glass, stainless steel, aluminium and teûon. The columns are either U shaped or coiled. The length of the column is in general 1.8 m with the dia of 2 mm.

Detectors: The detector is the most important component of GLC. The selection of the detector depends upon the type of compound, sensitivity and selectivity required in the analysis.

Thermal Conductivity Detector (TCD): It gives response to all compounds but it is not very sensitive.

Flame Ionization Detector (FID): It is a flame emission detector; the vaporised organic compound is burnt at the jet and ionization of organic molecules is measured. The response is dependent on the formation of ionic species. The sensitivity of FID detector is – 10^{-9}g consequently the detector is insensitive to c-cl, c-0, c-n and c-s and inorganic compounds.

Alkali Flame Ionization Detector (AFID): In this detector, the ionization processes at ûame are modiûed by the presence of alkali metal.

Nitrogen Phosphorous Detector (NPD): Basically it is an improvement over AFID

Electron Capture Detector (ECD): ECD is extremely sensitive to halogen compounds. This detector contains a beta radiation source such as tritium or ^{63}Ni which emits electrons and when an ionisable gas (N_2 or Argon) passes, it is ionized and slow electrons are produced. The type of detector is widely used for the detection of chlorinated pesticides uptopico gram levels.

GLC is the best technique for the analysis of pesticide residues because of its excellence sensitivity. It has found wider use in determining the trace amounts of pesticides.

High Performance Liquid Chromatography

High performance liquid chromatography (HPLC) or high-speed liquid chromatography (HSLC) or high-pressure liquid chromatography (HPLC) or high sensitivity liquid chromatography (HSLC) is the names used for this new Liquid chromatography. The generally accepted name is now high performance liquid chromatography (HPLC).

Solvent Reservoir

A reservoir should provide for solvent degassing and be of sufficient volume to allow for several analyses. The reservoir portion of the system can be described as limited or unlimited dependent on the pumping system.

Pumping Systems

The most important and most expensive part of the HPLC is the pump. Fluid flow essential for any HPLC separation is accomplished with a high-pressure pump. The high pressure is needed to overcome the resistance to fluid flow arising from the micro porous stationary phase packed in a narrow bore column. A desirable pump will provide constant flow at different rates with minimal pulsations at low and high pressures and be adaptable to gradient operation. There are two major design variations in HPLC pumps and these variations depend on the method of maintaining the fluid flow. Continues displacement (syringe) pump forces solvent from a single large reservoir into the column. This approach offers a smooth and pulse less solvent flow and is useful in operations requiring high detector sensitivity. Here periodic shut-down of the system is required to refill the reservoir, as the amount of fluid flow during pump stroke depends entirely on the size of the reservoir. The intermittent displacement pump avoids this refill problem by repeated refilling and pumping from a small chamber. A large external reservoir feeds this chamber, so shutdown for refilling is not necessary. During the refill stroke of this pump, no solvent flows to the column. The regular refill strokes impart a pulse like characteristic to the solvent flow. This pulsation appears at the detector output as tiny fluctuations on the chromatogram base line. Three different approaches are made to eliminate or reduce this flow noise.

- Use of baffle between the pump and the injector. A type of baffle commonly used contains a column of air that compresses and decompresses with full stroke cycle.
- Use of reciprocal pistons on the pump. These pistons are mechanically connected so that one piston is always pumping solvent into column while the other piston is ruling for the next stroke. This is widely used though the pumps are more complex and increasingly expensive.

- A third approach is with a single piston pump. At the end of a pump stoke, the pump speed is rapidly increased to refill the pump chamber. This rapid speed is maintained through the refill stroke by electronic device, which carefully monitor the flow rate, keeping it within set limits.

A HPLC system is termed isocratic whenever the solvent does not change composition during a given run. Most routine separations are generally isocraticalt. On certain occasions change in the mobile phase composition is desired during a run and is called as gradient programming. Solvent gradients can be formed two ways.

Low pressure gradient system: The eluents (solvents) are mixed before being pressurized by the pump. It employs a proportionating value for changing over the eluents flow paths. This is quite simple and inexpensive.

High-pressure gradient system: The eluents are mixed after being pressurized by the pump. It uses two or three pumps for pressurizing each eluents separately.

Mobile Phase

It should be free of particulates. Clogging of filters and column frits could disable the system. The pumps and columns are potentially sensitive to particulates use of 0.2-micron micro porous filter and/ or a 'guard' column just before the analytical column.

The removal of dissolved gases in the solvent is one other consideration. Under high pressure they remain dissolved and present less problem. However, when exiting the column the pressure drops dramatically, frequently allowing dissolved gases to form tiny bubbles. When these bubbles pass through the detector cell, signals to the recorder will become quite erratic.

Since the mobile phase governs solute-stationary phase interactions, its choice is critical. Strong acids (particularly halide acids), bases and halide solutions should be avoided. These corrode many system components and cause spurious results. Basic mobile phases should not be used with silica gels and for most bonded phases, because the bonding can be broken under alkaline conditions.

Chemical purity of solvent is another important factor. Since large volumes of solvent are pumped through the column trace solvent impurities can easily concentrate in the column and eventually produce detrimental effects. Only spectrograde or HPLC grade solvents are to be used.

Solvent compatibility with the detector is also important. All aromatic solvents are incompatible with UV and fluorescence detectors. Solvent with high viscosity reduces solute diffusion co-efficient; thus reducing mass transfer between solvent and stationary phase. Also, higher pump pressures are required. The mobile phase viscosity should be less than 0.5 centipoises.

Flow Rate

Slower the solvent flow rate, the better the separation. A typical flow rate for analytical separations is 0.5 to 2.0 ml/min. Pressures higher than 4000 psi are not normally encountered at analytical flow rate.

Sample Injectors

The injector is located between the column and the high-pressure solvent pump. Injections may be made using a syringe at high pressure, under stop flow conditions or using a sample loop. For best performance, the sample is to be placed at the head of the column in a narrow band.

Septum Injection

This is similar to the syringe injection in GC. Using high-pressure syringe, injections can be made through an inlet septum, which is a resilient seal at pressures upto 1500 psi. The septum must be compatible with the solvent and to withstand high backpressures. It allows variable volume of sample loading in narrow bands at the column head. The main drawback is the pressure limitation (1500 psi) and accumulation of small portions of the septum dislodged by the needle at the column head. This result in decreased column efficiency and/or plugging and excessive back pressure.

Stop Flow

This is essentially the same as direct syringe injection with the exception that solvent flow is stopped. When the solvent flow is stopped, the top fitting on the column is removed and the sample placed directly into the stationary phase. The column is then recapped and chromatography continues when the mobile phase flow is restarted. This is quite inexpensive and absolutely reliable. The resulting drop in pressure makes injection easy, since backpressure is eliminated. Column efficiency is little affected, since diffusion in liquids is slow. This approach is cumbersome to apply. Further the placement of materials directly into the delicate and expensive stationary phase is not recommended.

Loop Injection

This is by far the best and also is the most widely used. This requires a multiple port and a rotary switching valve. The solute mixture is first loaded into a fixed volume loop. Using the switching valve, this loop is switched into the path of the flowing mobile phase emptying the loop contents onto the column. The fixed volume loop provides reproducible injections at high pressure. By changing the size of the sample loop, sample volume can be selected.

Column

Columns are open tubular (precision-bore) stainless steel or glass (500 psi pressure limit) of 10-30 cm length with 10μ or less packing materials (stationary phase). The internal diameter (2-6 mm) depends on the packing, with the larger diameters used with the smaller particle size. Smooth internal walls, free from all surface defects are essential for packing the stationary phase evenly. Caps in the packing material can result in loss of column efficiency due to channeling (longitudinal cracks in the packing).

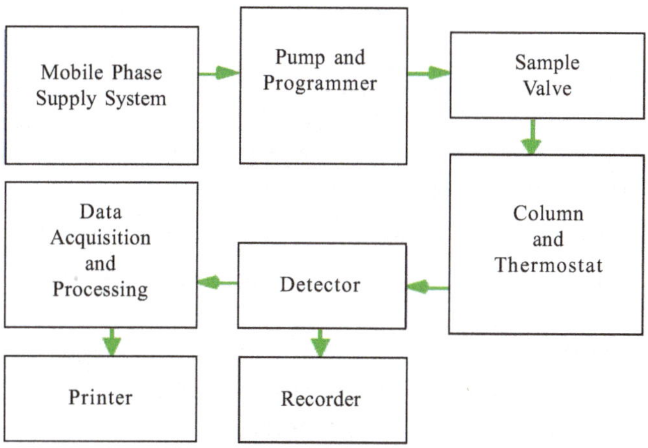

Schematic diagram of basic HPLC

Detectors

The HPLC detector provides electrical output signals proportional to the column eluent concentration and this electrical signal provide quantification. Though number of detectors like refractive index, flame ionisation, polarographic is used, only spectrophotometric and fluorescence detectors are of major value in residue analysis of herbicides with some extent the polarographic detector also be of value in herbicide analysis.

Table 2.8.10: Absorbing UV Chromophores

Chromophore	^Max	Absorbing Strength
Acetylide	177	Medium
Aldehyde	210	Strong
Azido	120	Medium
Amine	195	Weak
Ether	185	Weak
Ethylene	190	Medium
Ketane	190	Weak
Nitrile	225	Weak
Nitro	210	Strong
Oxime	190	Medium
Thiol	195	Weak
Thioketone	205	Strong
Thioether	194	Medium
Conj. ring	Varies	Strong

Strength is related to absorption coefficient

Weak = 0-4000; medium = 4000-8000; Strong = more than 8000

(*Source:* Instrumental methods of Analysis Willard, Merritt and Dean (1965).

Other Detectors are

Electrochemical detector, Electron capture detectors, Reaction detector, Transport detectors, HPLC mass spectrometric interfaces

Recorder

Recording the output of nearly all detection systems is easily accomplished with a variety of automatic peak integration and data reduction system.

Quantitative Determination of Residues

GC/LC is the most powerful analytical tool for quantitative analysis and is capable of giving accurate results. A typical chromatogram is enclosed to calculate the residue content in the sample.

$$\text{Residue in ppm} = \frac{\text{Weight of the standard}}{\text{Peak height of the standard}} \times \frac{\text{Peak height of the sample}}{\text{Weight of the sampe}} \times \text{Dilution Factor}$$

2.9

Herbicide Resistant Crops and Weeds

Resistance

The intensive and continuous use of herbicides over the last five decades has resulted in the development and evolution of weeds resistant to the normally phytotoxic chemicals. Although herbicide resistance was reported as early as 1957 against 2, 4-D from Hawaii the first confirmed report of herbicide resistance was against triazine herbicide in common groundsel (*Senecio vulgaris*) was reported in 1968 from U.S.A. Over the years, more than 150 weed species have been identified as having developed biotypes resistant to at least one or more herbicide classes. Resistance to at least 15 classes of herbicides have been noted. Moreover, the area of land infested with resistant weeds is increasing rapidly. Reports of evolution of newly resistant biotypes continue to emanate from all over the world, including developing countries, e.g., India, where extensive use of herbicides in grain crops began only in the late 1970s. These reports indicate that herbicide resistance problems are accelerating and, consequently, management of weeds is becoming increasingly more difficult and complex.

The phenomenon of herbicide resistance in plants, especially in weeds, is analogous to the evolution of changes that arise in insects and pathogens in response to the continuous use of insecticides and fungicides. Herbicide resistance is the result of selection for traits that allow weed species to survive specific herbicide usage, which would otherwise cause mortality. Herbicide resistance contributes a trait of resistance within the plant species to a herbicide and resistant weeds are those plant species that express the genetic variation required to evolve mechanisms to escape control. Like other crop pests (insects, pathogens, etc.) with reference to pesticides (insecticides, fungicides, etc.), herbicide-resistant weeds are the result of intensive selection pressure in weed populations.

Generally, resistance is defined as the ability of a plant species to withstand the phytotoxicity of a chemical. This is the trait normally found in crop plants, as against weeds, thus forming the basis for herbicide selectivity. Crop resistance to a herbicide is often different from the one that exists in the phenomenon of herbicide resistance in weeds.

Herbicide resistance in the context of weeds is defined as a characteristic of the weed species to withstand a herbicide dosage substantially higher than the wild type of the same plant species can withstand. It is the inherent or acquired ability to withstand the dosage of herbicide normally used for satisfactory weed control. Resistant weed populations are unharmed by a concentration of herbicide that completely kills all unselected populations. This herbicide concentration is the one that in field practice produces an effective kill.

Herbicide susceptibility represents the other end of the scale, encompassing natural variability in sensitivity to herbicides within the unselected populations. Tolerance, on the other hand, indicates reduced susceptibility (or conversely enhanced resistance) which may sometimes result from selection by herbicides. It is a response rather less than 100% survival accompanied by zero reduction in growth at herbicide concentrations, which normally give 100% Kill of susceptible biotypes. Tolerance may also be considered the natural or normal variability of response to-herbicides that exists within a species and can easily and quickly evolve.

There is no clear distinction between highly tolerant and resistant biotypes, and low levels of tolerance usually merge into susceptibility. The word resistance used encompasses both tolerance and resistance to herbicides in plant species and biotypes.

Development of Herbicide-resistant Crops

Herbicides are an integral part of modern agriculture because they provide cost-effective increase in agricultural productivity. These have traditionally been discovered by screening novel compounds in a series of tests, including greenhouse and field testing, toxicological testing, etc., spreading over several years. These tests are done to meet various biological, environmental, toxicological and economic criteria, so that the new compounds can compete with the many excellent existing compounds.

Selectivity is a function of the physicochemical properties of a compound and of thebiochemical interactions of the compound with the crop and weeds. In developing selective herbicides, the chemical companies resort to a bio rational approach in which the pathways of enzymatic degradation of herbicides, specific to the crop, are studied. The herbicidal moiety is then modified by adding a chemical group that will be recognized by those enzymes, thus aiding in the herbicide degradation by the crop. A number of important classes of herbicides such as triazines, sulfonylureas and imidazolinones have been developed by this approach. Another bio rational approach is to make current herbicides more effective by adding synergists (safeners or another herbicide) that either prevent their breakdown or prevent detoxiûcation of toxic products in some weeds.

In the past 15 years, biotechnologists in industry and the public sector have been resorting to another approach by making susceptible crops genetically and biochemically amenable to herbicides by conferring resistance. This genetic approach could remove a major factor in determining the choice of herbicides available for use by farmers. It could allow for the wider use of more effective herbicides with broader weed-control spectrums. In addition, compounds that are effective at low application rates have shorter persistence in the soil and also more favourable toxicological properties that might become more generally useful if the constraints of crop selectivity were removed. These possibilities would extend beyond major crops to minor crops that have not yet been benefited from effective herbicides. The introduction of herbicide resistance into crops could give the farmer greater ûexibility in choosing crops for rotational or multiple cropping systems. Thus, the combination of chemical approach and genetic approach could provide the farmer a broad range of crop management options for more effective weed control.

Biotechnological Approaches

Herbicide-resistant crops may be developed by using the following biotechnological approaches.

Plant Breeding and Selection

The plant breeding approach involves crossing, backcrossing and field testing before resistance is transferred into the varieties of choice. It requires two generations per year and at least five years of work before a herbicide-resistant crop variety can be field tested and released. Metribuzin resistance in soybean is inherited by a single dominant gene and was transferred to an important but sensitive soybean variety. Similarly, paraquat tolerance in *Lolium perenne* (perennial ryegrass), polygenically inherited, has been used to breed resistance in perennial ryegrass. The resultant material allows the use of low rates of paraquat in establishing stands of perennial ryegrass in pastures in England.

Many weed species show maternally inherited triazine resistance. Some of these weeds are sufficiently related to crops to allow transfer of the trait into related crop varieties. For example, transferring the trait from *Setaria viridis* (green foxtail) into the crop *Setaria italica* (foxtail millet) was moderately successful. The yields of the resultant crop varieties were considerably lower than the original variety. This loss could be alleviated by further hybridization but so far the resultant hybrids have always yielded less than the reciprocal hybrids whose offspring were triazine-sensitive. The loss of yield is probably due to the low fitness that is usually inherent in the chloroplast genome mutation conferring triazine resistance. The triazines are good candidates for developing crop resistance as they are the least expensive herbicides. However, there is the possibility of evolution of more triazine-resistant weeds if triazine herbicides are used continuously without rotating with non-photosystem II-inhibiting herbicides.

Resistance can be transferred by selection at the whole plant level. Seeds may be mutagenised to isolate resistant genes for genetic engineering, as in the case of *Arabidopsis thaliana*. Another plant in which the whole plant selection approach is used is the fern *Ceratopteris* spp. Selection pressure is applied on the photo autotrophic haploid gametophytes, allowing easier selection of recessive mutants for resistance to herbicides such as paraquat and acifluorfen. Paraquat resistance is monogenically inherited. These mutants can be used, after analysis of fitness, for selecting the resistant genes for genetic engineering. Using similar methods, large numbers of mutation-treated soybean seeds were selected for chlorsulfuron tolerance. However, none of the mutants had target site resistance at the level of ALS.

Tissue Culture

Cell cultures have short generation times (doubling in 3-5d), 10^5-10^6 cells ml^{-1}, with each cell theoretically representing a whole plant. This allows the possibility of very uniform treatment with herbicides at high selection pressures. Thus, cell cultures have many advantages over field cultures for selection of resistance. However, many species cannot be regenerated from tissue cultures and only some varieties of some species can be regenerated, although many recalcitrant crop species including maize, wheat and soybean have been regenerated from cell cultures in the recent past. In tissue culture, selection can be done using non-photosynthetic cells, green tissues and protoplast fusion transfer.

Genetic Engineering of Herbicide Resistance in Crops

Genetic engineering in crops for herbicide resistance has become a major area of weed research for over a decade. The process of genetic engineering involves isolating the gene from resistant plants or microorganisms (fungi), inserting it into plasmid and then into a vector system, introducing it into plant cultures and regenerating the cultures as a newly resistant crop plant. The main advantage of the genetic engineering approach is that once the gene is successfully engineered into a plasmid vector system, it can be used to transform many different plant species. Besides Solanaceous species, which have been used as models, the *Agro bacterium* infection systems (with model marker genes) have been used to successfully transform soybean, cotton, sunflower and maize. Many species have been transformed by incubating protoplasts with plasmid DNA and pushing it through the cell membrane with polyethylene glycol, or electrophorating it through the membrane with a high voltage pulse or even pelting cells with DNA or micro projectiles.

Genetically Modified Crops and Herbicide Tolerance

The most compelling case for biotechnology, and more speciûcally GM crops, aretheir capability to contribute to: increasing crop productivity and thus contribute to global food, feed and ûber security; conserving biodiversity, as a land saving technology capable of higher productivity; more efficient use of external inputs and thus a more sustainable agriculture and environment; increasing stability of production to lessen suffering during famines due to abiotic and biotic stresses; to the improvement of economic and social beneûts and the alleviation of abject poverty in developing countries.

Between 1996 and 2000, the area of genetically modified (GM) crops increased from 44 million ha. Soybean, maize, cotton and oil seed rape are four of the major crops modified and herbicide tolerance and insect protection are the two main traits introduced. Glyphosate and glufosinate-ammonium are the two main herbicides, for which crops have been modified for herbicide tolerance (HT), Herbicide tolerant soybean is currently the dominant transgenic crop globally. While there is evidence to indicate that the introduction of HT soybean will reduce herbicide use by up to 10%. If a 10% reduction in pesticide use is assumed then the use of herbicide tolerant soybeans reduced pesticide use by 2.9 million kg formulated product. The area of HT canola is small in comparison to soybean, in addition to finding that the introduction of HT canola increased yields by about 10%, it was reported that transgenic crops required less herbicide than conventional crops. The total amount of herbicide used was reduced by 1.5 million kg in I997 and by 6.0 million kg of formulated product in 2000. Herbicide use in cotton is expected to decline with the adoption of HT cotton varieties. Application rates for conventional varieties vary from 4.9 to 8.0 kg formulated product/ha compared with 2.5 to 4.0 kg/ha for glyphosate tolerant GM varieties. Thus with 2.1 million ha grown the decrease in pesticide use associated with the introduction of HT cotton is estimated at 5.3 million kg formulated product in 2000. The area of HT maize grown globally is 2.1 million ha, the use of herbicide tolerant maize has on average reduced herbicide use by 30% (0.69 kg/ha). This is equivalent to a reduction of 1.5 million kg formulated product/ year.

With India growing a GM crop for the first time in 2002 the three most populous countries in Asia-China, India and Indonesia with 2.5 billion total people, are all now commercializing GM crops.

Globally, transgenic soybean occupied 36.5 million hectares in 2002, with transgenic corn in second place at 12.4 million hectares, transgenic cotton in third place at 6.8 million hectares and canola at 3.0 million hectares. Herbicide tolerant soybean continued to be the dominant transgenic crop grown commercially. Herbicide tolerant maize on 2.5 million hectares (4%).Bt cotton on 2.4 million hectares (4%); herbicide tolerant cotton on 2.2 million hectares(4%); Bt/herbicide tolerant cotton on 2.2 million hectares (4%); and Bt/herbicide tolerant maize on 2.2 million hectares (4%).

On a global basis GM technology has reduced pesticide use (includes herbicide), with thesize of the reduction varying between crops and the introduced trait. It is estimated that the use of GM soybean, oil seed rape, cotton and maize varieties modiûed for herbicide tolerance and insect protected GM varieties of cotton reduced pesticide use by a total of 22.3 million kg of formulated product in the year 2000. Globally, the principal GM crops were: GM soybean, GM corn, transgenic cotton and GM canola.

Table 2.9.1: Dominant GM crops

Crop	Trait	Area (in ha)	% total GM crop area
Soybean	Herbicide resistance	33.3	63
Maize	Total all traits	9.8	18
	Insect resistance (Bt)	5.9	11
	Herbicide resistance	2.1	4
	Stacked Bt /herbicide resistance	1.8	3
	Total all traits	6.8	14
	Herbicide resistance	2.5	5
Cotton	Insect resistance (Bt)	1.9	4
	Stacked BT/ herbicide resistance 2.4	2.4	5
Oilseed rape	Herbicide resistance	2.7	5

India, the third largest cotton growing country in the world, has approved the commercial application of insect-resistant GM cotton in 2002 (James, 2002), and is expected to grow over0.15 million ha of GM cotton and further.

Genetic Engineering

Genetic engineering, recombinant DNA technology, genetic modiûcation manipulation (GM)and gene splicing are terms that apply to the direct manipulation of organisms' genes. Genetic engineering is different from traditional breeding, where the organisms' genes are manipulated indirectly. Genetic engineering rises the techniques of molecular cloning and transformation to alter the structure and characteristics of genes directly. It is used for improving crop technology.

MAPPING OF DNA

DNA is a double-stranded helix. The two strands are connected by the chemical bases A, C, G and T.A. pairs with T; G pairs with C. A gene is a segment of DNA that has a specific sequence of these chemical base pairs

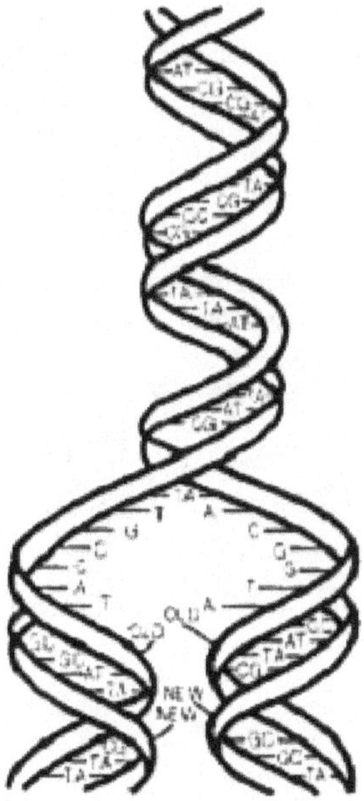

Protein synthlzesis - Special enzymes copy a singe DNA strand to make a single messenger RNA (mRN) strand in which U replaces T as a base.

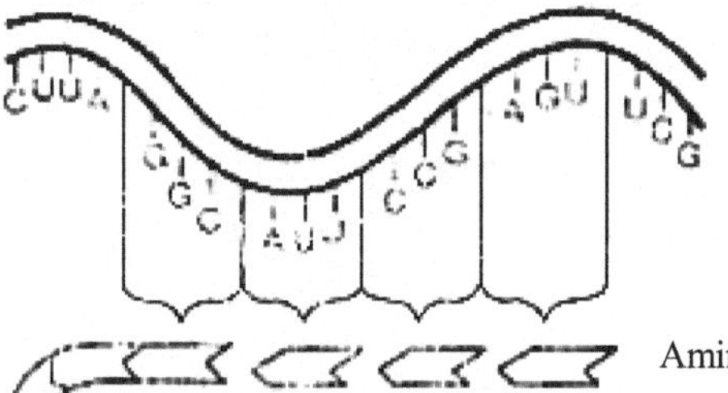

Amino Acid

The mRNA moves out of the cell nucleus and into the cytoplasm where each three-letter sequence is translate into one amino acid. Chains of amino acids make up the proteins that compose all living things

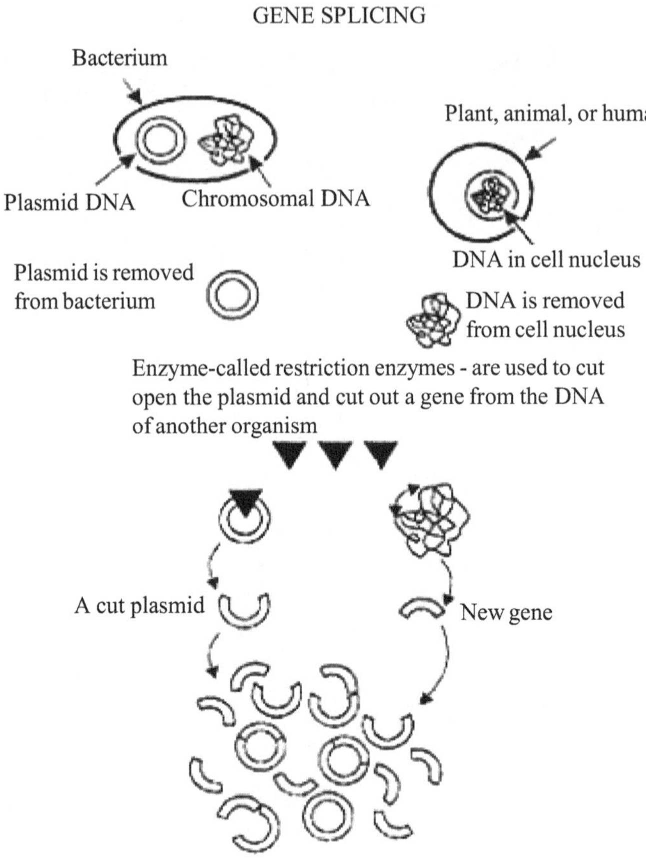

Fig. 2.9.2: The cut ends of the plasmids and the cut ends of the new genes are chemically "sticky", so they will attach to each other -recombine -to form a new loop containing the inserted gene. This technique is called gene splicing in recombinant DNA.

The essential processes for accomplishing Genetic engineering are

1. Isolation of the gene of interest
2. Insertions of the genes into a transfer vector
3. Transfer of the vector to the organism to be modiûed
4. Transformation of the cells of the organisms
5. Selection of the genetically modiûed organism from those that have not been successfully modified.

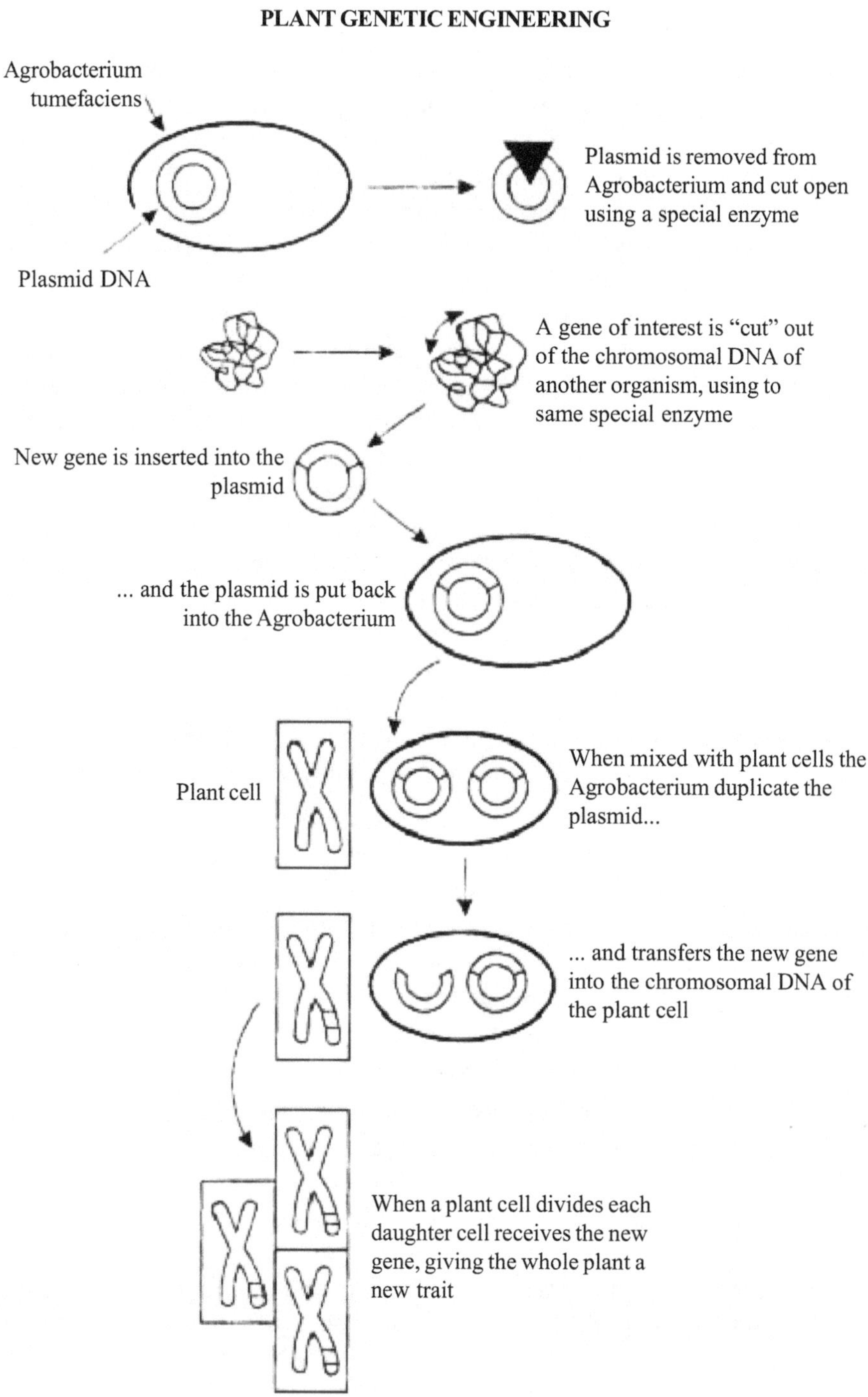

Fig. 2.9.3: Examples of how plants are genetically engineered (Subramaniam *et al.*, 2011)

Herbicide Tolerance Through Engineering of Plants

Application of agro bacterium mediated gene transfer in agriculture is one of the ways to develop tolerance. The exiting progress has been made in applying to the transfer of agriculturally useful genes, which includes genes for insect and disease resistance and tolerance to safer herbicides. Agro bacterium mediated gene transfer is the only method of non sexual gene transfer for which there are now practical and useful examples that are being tested and that are serious candidates for use in agriculture. An early goal in the use of recombinant DNA for crop improvement has been to engineer bacterial or plant genes encoding enzymes that make crop plants tolerant to broad spectrum environmentally safer herbicides. One successful strategy has been to transfer genes for an enzyme that complements the plant enzyme whose action is blocked by the herbicides. This has been done by engineering a bacterial gene so that its enzyme product is insensitive to the herbicide and then transferring in to the plant. Alternatively, the plant produces a large amount of its own enzyme making plants that can survive in the presence of the herbicide. Another strategy is to engineer pants to express genes for enzymes that chemically detoxify the herbicide.

Herbicide Resistance in Weeds

Herbicide resistance is the inherent ability of a species to survive and reproduce following exposure to a dose of herbicide normally lethal to its wild type.

Over the past years, repeated use on the same ûeld site of herbicides with similar modes of action has imposed selection for increased resistance within species that formerly had been susceptible. Herbicide resistance has become well known in scientiûc and agriculture communities around the world since the discovery and report of triazine resistance in common groundsel in 1970. As resistance is generally the consequence of using a single herbicide repeatedly, any proactive or reactive approach should take an opposite view: use a diversity of methods to avoid repetition as much as possible.

One result of modern agriculture and the reliance on herbicides is the emergence of populations of weeds which are resistant to products designed to control them. All natural weed populations regardless of the application of any weed killer probably contain individual plants (biotypes) which are resistant to herbicides. Repeated use of herbicide will expose the weed population to a "selection pressure" which may lead to an increase in the number of surviving resistant individuals in the population. As a consequence, the resistant weed population may increase to the point that adequate weed control cannot be achieved by the application of that herbicide. The ûrst case of herbicide resistance in weeds was identiûed in 1964. Currently there are recorded more than 150 resistant grass and broadleaf weed biotypes in about 50 countries world-wide (Heap I997). In spite of this seemingly dramatic development, no herbicides have been lost to agriculture; they are today, and will remain, an integral part of food production through their effective use in combination with other weed control practices.

Worldwide, herbicides remain the most efficient technology for large-scale weed control. Therefore, the widespread evolution of herbicide resistance in weed populations within intensive crop production systems is a major threat to the sustainability and profitability of cropping systems. The introduction of new herbicides and herbicide modes of action to replace those herbicides failing due to resistance is essential for weed management. To date, resistance to 19 classes of herbicides has been reported in 330 biotypes of 189 species (113 dicots, 76 monocots) worldwide. Sixteen prominent weed species have developed resistance to glyphosate alone.

Management of Herbicide Resistant Weeds

Use of Transgenic Herbicide Resistant Crops

Herbicides are either selective in that they kill weeds without injuring certain crops or nonselective in that they kill most annual weed and crop species. A good selective herbicide can be used in particular crops but will injure other crop species. In the herbicide discovery process, crop selectivity is largely the result of the skill of the inventive scientist. Crop selectivity is usually due to the ability of the crop to detoxify the herbicide at faster rate than that which occurs insusceptible weed species. In contrast, non selective herbicides (glufosinate, glyphosate andparaquat) are lethal to wide spectrum of annual weed and crop plants and there is little or no metabolism of these herbicides in either crops or weeds. The ideal herbicide would control awide spectrum of weed species, be harmless to crop species of choice and have desirable environmental characteristics. Recent bio technological advances in plant sciences provide agenetic means of endowing crop species with resistance to herbicides of choice. Transgenic crop resistance to herbicide provides useful tools for the grower. Although in principle it is feasible to select or engineer plants for resistance to any herbicide, the reality is that there are relatively few forms of herbicide - resistant crops that are either in development or have been commercialized. Some examples are :

1. **Glyphosate resistance (Group M)**: Glyphosate is a post emergence, non selective herbicide. Research at Monsanto and elsewhere over the past decade has led to the identiûcationof glyphosate resistance genes and their subsequent successful expression crop species where resistance to glyphosate is endowed. Glyphosate-resistant soybean has a high level

of tolerance to twice the average use rate of glyphosate and the applications of 0.63 kg acid equivalent, which provides effective season long, broad spectrum weed control in soybean. Transformed cotton plants, containing the resistant enzyme, also show high levels of resistance, with no yieldpenalities. Two glyphosate resistant canola lines are being developed primarily for western Canada. By comparison, obtaining sufficient levels of resistance in com is proving more difficult,but lead lines with both resistant EPSP synthase and GOX gene are being evaluated.

2. **Glufosinate Resistance (Group N)**: The herbicides glufosinate and bialaphosare inhibitors of glutamine synthetase, the only plant enzyme that can detoxify ammonia released by nitrate reduction, photo respiration and amino acid accumulation. Inhibition of this enzyme results in an accumulation of ammonia which leads to death of the plant. Both products are used as nonselective herbicides and are active following post emergent applications on a wide range of grass and broad leaved weeds. Research has been focused on developing glufosinate resistance in four major crops-canola, maize, soybean and sugarbeet. Application of glufosinate at 300-600 g/haconsistently provided good weed control and proved safe to the crop.

3. **Bromoxynil Resistance (Group C)**: Bromoxynil is a PS II inhibitor that inhibits electron transport. A bromosynil resistant cotton has been developed by Calgene in conjuctionwith Rhone Poulenc and there has also been some interest in transforming other species with resistance to this herbicide. Transgenic cotton containing the bxn gene can efficiently metabolize-bromoxynil to 3,5 dibromo-4-hydroxybenzoic acid and other products.

4. **ALS-Inhibitor resistance (Group B)**: ALS inhibitor resistant cultivars of canola have been generated by selection in cell and microspore culture or by agro bacterium–mediated gene transfer. Mass selection of mutagenized seed has been used to generate ALS inhibitor-resistant soybean. Canola expressing a herbicide resistant ALS is highly resistant to sulfonylurea andtnazlopynmidine herbicides and has similar maturity, seed yield and oil content as standard/ cultivars. The imidazolinonc-resistant maize lines display varying/levels of-resistance to ALSinhibitors and also have no yield penalities.

Merits and Demerits of Transgenic Herbicide Resistant Crops

A list of the perceived benefits that may arise through the use of such crops are:

- Selective control of difficult weeds
- Increased use of environmentally benign herbicides
- Reduced herbicide rates
- Additional weed control options available to the grower '
- Greater timing ûexibility for growers
- Reduced soil erosion
- Control of herbicide resistant weeds
- Economic advantages to growers and consumers
- Improved returns for herbicide discovering industry

In contrast, the perceived risks associated with herbicide resistant crops can be summarized as :

- Greater use and reliance on a few herbicides
- Unfavourable environmental impact
- Gene transgression to wild or cultivated plants
- Pleiotropic effects of transgenes
- Development of volunteer weed problems
- Development of herbicide resistant weeds
- Public and consumer acceptability of transgenic plants and their products
- Reduced bio diversity
- Development of monopolistic chemical/seed companies.

Mechanisms of Herbicide Resistance in Weeds

Mechanisms of herbicide resistance include altered site of action, overproduction of the site of action, enhanced herbicide metabolism, decreased herbicide absorption and translocation, and herbicide sequestration. The most common mechanisms that cause herbicide resistance in weeds are altered site of action and enhanced herbicide metabolism. Site of action is the speciûc process within a plant that a particular herbicide or herbicide group inhibits. Alterations in the site of action that prevent the herbicide from binding are the most common mechanism of resistance. Enhanced herbicide metabolism is the superior ability of plants to convert the herbicide molecule to a form that is no longer toxic, whereas susceptible species are unable to detoxify the herbicide at the same rate as resistant species.

Mechanisms of herbicide resistance can be broadly grouped into two categories

1. **Exclusionary resistance:** Those that exclude the herbicide molecule from the site in plants where uVy induce toxic response. In exclusionary resistance mechanism the herbicide is excluded from the site of action in many ways.

a) **Differential herbicide uptake**: In resistant biotypes the herbicides are not taken up readily due to morphological uniqueness like overproduction of waxes, reduced leaf area etc.

b) **Differential translocation**: In resistant biotypes the apoplastic (cell wall, xylem) and symplastic (plasma lemma, phloem) transport of herbicide is reduced due to differentmodifications.

c) **Compertmentation**: Herbicides are sequestered in many locations before it reaches the site of action, e.g. some lipophilic herbicide may become immobilized by partitioning into lipid rich glands or oil bodies.

d) **Metabolic detoxification**: Herbicide is detoxified before it reaches the site of action at a rate sufficiently rapid that the plant is not killed. The biochemical that detoxifies herbicides can be grouped into four major categories: oxidation, reduction, hydrolysis, and conjugation. Three enzyme systems are known to be involved in resistance due to increased herbicide detoxification. Resistance to atrazine in some population of *Abutilon theophrasti* is due to increased activity of glutathione-s-transferase that detoxiûes atrazine.

Site of Action Resistance

Those that render specific site of herbicide action resistant

a) **Altered site of action**: Site of action is altered in such a way that it is no longer susceptible to the herbicide e.g. In *Lactuca sativa* biotypes which are resistant to sulfonyl urea herbicides, the ALS enzyme which is the site of action of herbicide ismodiûed in such a way that herbicide can no longer bind with the enzyme and inactivate it. This target site based resistance is usually associated with resistance involving altered binding of herbicide to their target protein. This results from a single nucleotide change (mutation) in the gene encoding the protein to which the herbicide normally binds. This change the amino acid sequence of the protein and reduces or destroys the ability of the herbicide to interact with the protein and at the same time do not incapacitate the normal functioning of the enzyme so that the enzyme functions normally in the presence of the herbicide. However these mutations conferring herbicide resistance may cause changes in other seemingly unrelated physiological processes.

Usually these changes adversely affect the biological fitness of the resistant biotypes.For example, in triazine resistant biotypes the mutation in the plastoquinone binding D1 protein of PSII results in reduced photosynthetic efficiency also seeds of some of these resistant weeds biotypes exhibit poor germination as compared to susceptible biotypes. But in some resistant biotypes, as found in *Kochias coparia*, the mutation conferring resistance to sulfonylurea herbicides will concomitantly reduce or abolish acetolactate synthase sensitivity to normal feedback inhibition patters, resulting in elevated levels of branch chain amino acids available for cell division and growth during early germination. Hence sulfonyl urea resistant biotypes of this species exhibit a rapid germination even at lower temperature compared to their susceptible counterpart.

b) **Site of action overproduction**: This causes the dilution effect of the herbicide. Here the site of action is overproduced so that the herbicide at its normal rate of application will not be able to inactivate the entire enzyme produced. Thus the enzyme spared by the herbicide will carry on the normal plant metabolic activities.

Types of Resistance

Cross- or multiple resistance: Resistant weeds can survive the application of herbicides other than the herbicide to which they have developed resistance (i.e., the selecting herbicide).In such cases, resistant weeds are considered to have cross- or multiple resistances. Cross-resistance occurs when one resistance mechanism (e.g., enhanced herbicide metabolism) allows the plant to withstand herbicides from different chemical classes. However, when a plant has multiple resistance it possesses two or more distinct resistance mechanisms (e.g., two or more altered sites of action), which allow the plant to resist herbicides from different chemical classes.

Negative cross resistance: On the contrary, in some weeds, resistance to one herbicide results in increased susceptibility to another herbicide or other abiotic factors, such as standard cultivation practices, and/or biotic factors, such as effects of insect pests or infection by viruses and fungi. This phenomenon is known as negative cross-resistance and could be exploited in some resistant weed species.

Factors Affecting Rate of Resistance Development

Resistance is not due to mutation caused by herbicides; rather it arises from the selection of natural mutation or small pre-existing population of resistant plants (selection pressure exerted by herbicides)

1. **Selection Pressure**: Selection pressure is one of the most important determinants of resistance development. Major factors contributing to selection pressure include the efficacy of the herbicide and the frequency of use. In general, selection pressure is a measure of the ability of an herbicide to differentiate between susceptible and resistant plants. Therefore, herbicides with higher efficacy in the control ofa speciûc weed population impose higher selection pressure for selecting the resistant individuals among that weed population Moreover, depending on the soil and herbicide chemical properties, herbicide residual activity could also be a major factor in the development of herbicide resistance. This means that the selection pressure on susceptible weeds from herbicides with longer residual activities would be higher than that from herbicides with shorter or no residual activities.
2. **Initial Frequency of Resistance Mutations**: Precise studies have not been performed on the initial frequency of resistance mutations in weeds. However, some estimates have shown that approximately one out of a billion seeds from a population could be resistant. While these mutations are rare, large weed populations increase the likelihood that they might be present.
3. **Gene Flow**: Pollen and seed movement are the major methods of gene flow between and within plant populations. Furthermore, the spread of resistance occurs more rapidly in cross pollinated species compared to self-pollinated species Therefore, resistance genes mostlikely arise in an area through mutation, and gene flow would facilitate the spread of the resistance genes among individuals within that area.
4. **Fitness**: Fitness is a measure of survival and ability of a given genotype (e.g., herbicide-resistant biotypes) to produce viable offspring in competition with the wild type (e.g., herbicide-susceptible biotypes). Although herbicide-resistant biotypes should be less ût than susceptible biotypes, depending on the mechanism of resistance and the environmental conditions, the fitness of resistant populations may vary. For example, under agricultural field conditions, triazine resistant biotypes have been shown to be lesser than the susceptible biotypes, but many studies have not been able to detect fitness penalty in biotypes resistant to AHAS inhibitors (e.g., Pursuit) under those conditions. The existence of fitness penalty under field conditions could be exploited for management of those resistant biotypes affected by it.
5. **Initial frequency of the resistant individuals**: If the initial frequency of the resistant individual is high in a natural weed population, then the resistance will surface more quickly than in a population where the

frequency of the resistant individual is low, provided weare continuously applying the herbicide to which the biotypes exhibit resistance.

6. **Weed seed residue in the soil seed bank**: For a species if the seed residue is more in the soil seed bank, appearance of resistance will be delayed due to continuing recruitment of susceptible individual from soil seed bank. That is, Nature will allow the resistant species toûourish only after major portion of the susceptible weed seeds have been exhausted from the soil. For this very reason the species that germinate readily from its propagules will develop resistance more quickly than those species whose propagules remain dormant in the soil.
7. **Hypersensitivity of weeds to a particular herbicide**: Because of hypersensitivity,with a single application the herbicide about 90-95% of the susceptible type is killed. So selection pressure will be high and resistance species evolve rapidly.

Herbicides

1. **Lack of rotation of the herbicides:** Continued application of the same herbicide or different herbicide with the same mode of action will create selection pressure and will allow resistant population to flourish.
2. **Herbicides with long residue period**: This results in continued suppression of susceptible biotypes for a longer period, thus allowing the resistant species to ûourish.
3. **Herbicides with highly specific mode of action**: If a herbicide has only onesite of action in weeds, then a biotype need to be different in that particular site to be resistant.So the evolution of resistance against such herbicides will be quicker than against herbicides having multiple site of action.

Strategies for Managing and Preventing Herbicide Resistant Weeds

The management practices must be primarily focused on reducing the selection pressure.

a) **Rotation of herbicides with different mode of action**. Use of the same herbicide of different herbicides with the same mode of action will exasperate the problem of resistant weeds. So adopt rotation of herbicides with different mode of action.

b) **Use of herbicide mixtures**: Herbicide mixtures are presently employed to broaden its spectrum of activity. But resistant management requires both the components of mixture control the same spectrum of weeds so that the weeds resistant to vulnerable herbicides will be destroyed by the mixing partner, or at least be rendered relatively comût compared to the wild type. Evolution of target-site resistance to both vulnerable and partner herbicide, though possible when mixtures are used, are much delayed. The following reasoning based on compound resistance has been used to support this supposition. If frequency of individual resistant to each components of a pesticide in a mixture is independent in susceptible species, then joint probability of evolution of co-resistance to both herbicide in one individual equals the product of the probabilities of resistance for each partner. Thus if a weed has a natural mutation frequency of 10-5 for resistance to vulnerable herbicide and 10-10 to mixing partner having different target site and if genes for resistance are inherited independently of each other, then the joint probability of resistance to both the herbicide in an individual will be 10-15 which is very rare.

Characteristics of Effective Mixing Partner

a) It must kill same spectrum of weeds as the vulnerable partner.

b) It must have a mode of action different from that of vulnerable partner.

c) Both must have some effectiveness in weed control: It may not be helpful, if at its rate used, the mixing partner kill 75% of the weeds and vulnerable kills 95% unless the 20% remaining are severely inhibited such that they have less reproductive capacity than wild type. Otherwise resistance could quickly evolve in remaining 20% of weeds.

d) Both components must have similar persistence: Otherwise there will be period when only the vulnerable one is present and since the weeds have many flushes of germination during a cropping season the target weeds will not be exposed to mixture.

e) The mixing partner should not be degraded in the same manner as vulnerable partner

f) It will be an added advantage if the mixing partner posses negative cross resistance i.ewhere individuals resistant to vulnerable herbicides are more susceptible than the wild type to the mixing partner.

c) **Use of herbicides when only necessary**: indiscriminate use of herbicide like pre-emergent application of herbicide must be avoided wherever there is an option for selective post-emergent herbicide. Adoption of herbicide resistant crops can also help us in this respect.

d) **Control of weed escapes and sanitation of equipment to prevent spread of resistantweeds:** Weed escapes must be prevented by adopting optimum dose, time and method of application of herbicides. Dissemination of resistant weed must be prevented.

e) **Use of herbicides with short residual life**: If we are using herbicides having long residual life then the selection pressure will be more. So use herbicides having short residual life. Also, if we are increasing the dose of herbicide the residual period will be high. So use the recommended dose.

f) **Scout the fields for resistant weeds**: Before and after herbicide spray take walk through the ûelds observing the weed flora. If, after the application of herbicide, you are running into a patch of weed escape, destroy it.

g) Adopt integrated weed control practices

h) **Adopt crop rotation**: Crop rotation usually means using diverse herbicide program making it difficult for resistant weed to increase.

Herbicide resistant weeds

Herbicide Group	Mode of Action	HRAC	Example Group	Total Herbicide
ALS inhibitors	Inhibition of acetolactatesynthase ALS (acetohydroxyacid synthase AIIAS)	B	Chlorotoluron	70
Photosystem II Inhibitors	Inhibition of photosynthesis at photosystem II	CI	Atrazine	63
ACCase Inhibitors	Inhibition of acetyl CoA carboxylase (ACCase)	A D	Diclofop methyl Paraquat	27 21
Bipyridiliums	Photosystem-I-electron diversion	O	2,4-D	21
Synthetic Auxins	Synthetic auxins (action like indoleacetic acid)	Q	Chlorotoluron	20
Ureas and amides	Inhibition of photosynthesis at photosystem II	Kl	Tnûuralin	10
Dinitroanilines	Micro tubule assembly inhibition	N	Triallate	6
Thiocarbamates and others	Inhibition oflipid synthesis - not AC Case inhibition	F3	Amitrole	4
Triazoles, ureas isoxazolidiones	Bleaching: Inhibition of carotenoid biosynthesis (unknown target)	G	Glyphosate	4
Glycines	Inhibition of EPSP synthase			
Chloroacetamides and others	Inhibition of cell division (Inhibition of very long chainfatty acids)	K3	Butachlor	2
Nitriles and others	Inhibition of photosynthesis atphotosystem	C5	Bromoxynil	1

Carotenoid biosynthesis inhibitors	IIB leaching: Inhibition of carotenoid biosynthesis at thephytoenedesaturase step (PDS)	Fl	Flurtamone	1
Mitosisinhibitors	Inhibition of mitosis/ microtubulc polymerization inhibitor	K2	Propham	1
Organoarsenicals	Unknown	Z	MSMA	1
Arylaminopropio -nic acids	Unknown	Z	Flamprop-methyl	1
Pyrazoliums	Unknown	Z	Difenzoquat	1

Total Number of Unique Herbicide Resistant Biotypes 254 (Heap, 2002)

3.1

Weed Shift and Its Management

Shifts or changes in the abundance and types of weeds within agricultural systems are common. These shifts can occur for a number of reasons and may result from cultural, mechanical, or chemical weed management strategies. The right combination of weeds and relying too heavily on any one practice is often the most common cause. Some of the concepts associated with weed species shifts are discussed below.

A weed shift is the change in the composition or relative frequencies of weeds in a weed population (all individuals of a single species in a defined area) or community (all plant populations in a defined area) in response to natural or human-made environmental changes in an agricultural system. Weed shifts occur when weed management practices do not control an entire weed community or population. Some species or biotypes are killed by (or susceptible to) the weed management practice, others are not affected by the management practice (tolerant or resistant), and still others do not encounter the management practice (dormant at application).

Those species that are not controlled can grow, reproduce, and increase in the community; resulting in a weed shift. Any cultural, physiological, biological, or chemical practice that modifies the growing environment without controlling all species equally can result in a weed shift. Weed shift may be due to introduction of new species or change in relative proportion of members of the existing community. Latter is the most common.

A. Species shift: Change in density of one or more species in field

B. Biotype shift: Change in density of biotypes within a species. These shifts often found noticed since biotypes typically are similar in appearance. Herbicide resistantbiotypes are the most frequently studied.

Causes: Each member of the weed community has different optimum growing conditions and responds differently to control practices. The community within a field is a direct result of current and past environment and management practices. The community is in constant flux as management practices change.

Weed management programs have the biggest impact on weed communities. Since no control tactic is equally effective against all members of a weed community, individuals (species/biotypes) that are best adapted to survive current tactics will increase, those less adapted will decrease – survival of the fittest.

Examples of weed shifts to various management practices

A. No-till: Increase in weeds with perennial and winter annual life cycles

B. Vegetable crops: Weeds with short life cycles (purslane, hairy galinsoga)

C. Low mowing height: Weeds that can survive with prostrate habit - large crabgrass,annual bluegrass, and dandelion

D. Grazing: Thistles (grazers avoid them due to spines), toxic plants

E. Hand weeding: Crop mimicry - In rice fields, barnyard grass biotypes with identical vegetative characteristics of rice, so weeders couldn't distinguish the weeds from the crop.

Documented Weed Shifts

Weed shifts have occurred throughout history as people have modified cropping systems and the environment for the production of desired crops. These shifts can be immediate or short-term fluctuations in response to transient changes in production systems, or they may represent long-term changes. Shifts occur within weed populations, weed communities, or concurrently within both, in response to single or multiple changes in agricultural systems. Weed shifts have occurred in response to changes in tillage, irrigation, fertility, crop rotation, and herbicide use practices. Examples of weed shifts that have occurred in recent history include the following:

Weed Shift Due to Alien Invasive Weeds

Alien invasive weeds which are non - native to the eco system cause harm to ecological, economic & human health. These plants are implicated in the decline of threatened & endangered species because they alter ecosystem processes, changes community structure & displace native species. About 18% of the Indian flora constitutes adventive aliens. *Chromolaena odorata, Lantana camara, Parthenium hysterophorus* & *Prosophis julifera* among terrestrial exotics have posed serious threat to the native flora.

Community Shift in Response to Herbicide use

In the Corn Belt and winter wheat areas of the western United States, changes in weed communities were noted within 10 years of the introduction of 2,4-D for the control of broadleaf weeds. In corn, summer annual grass species increased as broadleaf species were controlled. In wheat, winter annual grass species replaced broadleaf species as the predominant troublesome species. Continuous use of isoproturon for weed control in rice - wheat system has led to weed shifts from grassy to broad leaved weeds like *Rumex dentatus and Malva parviflora.* Due to continuous use of grass killers in rice, the weed flora is shifting to sedges like *Cyperus* sp., *Scripus* sp., *Fimbristylis* sp., *Elecaris* sp. With the continuous use of butachlor in rice, some tolerant weeds like *Ischaemum rugosum, Caesulia* sp., *Sphenoclea* sp. *etc.* have become dominated in rice fields. Similarly continuous use of butachlor 0.75 kg + 2, 4 - d EE 0.4 kg /ha (3 DAP) both during kharif and summer paved way for dominance of sedges (particularly *C. difformis* and *F. miliacea*) and grasses (*E. glabrascens*) during fourth summer crop of rice - rice system. *Phalaris minor* was the most predominant grassy weed in wheat crop. Clodinafop and sulfosulfuron were most common herbicides adopted for weed control. As these herbicides do not control Poa annual, its intensity is on the increase. *Medicago denticulata, Melilotus spp. and Lathyrus aphaca* in wheat fields are increasing due to continuous use of grassy weed killers like isoproturon. *Solanum nigrum*is not being controlled by the herbicides like sulfosulfuron, 2, 4-D or metsulfuron methyl. *Polypogon monspeliensis* was seen in high density in wheat crop having excess moisture content in the soil.

Community Shift in Response to Tillage Change

Changes from conventional to reduced tillage systems often causes weed community shifts that include increases in summer annual grasses and small-seeded summer annual broadleaves, winter annual, biennial, and perennial species, and decreases in large-seeded summer annual species. In maize - sunflower cropping system, when zero tillage is practiced for both the systems continuously, there was a shift in weed flora from broad - leaved weeds to grasses.

Community Shift as a Result of New Localized or Long-distance Introductions

Common lambs quarters, a weed believed to be native to Europe and Asia, is now found throughout much of the United States. In much of Pennsylvania, common lambs quarters has become predominant in the weed community. The shift occurred because the species grows aggressively, is difficult to control, and is a prolific seed producer. Long-distance dispersal has also resulted in the introduction of many noxious weeds to the United States, some of which have caused weed community shifts (including field bindweed in the western plains, leafy spurge in rangeland, and multiflora rose in pasture).

Population of Biotype Shifts in Response to Herbicide use (herbicide – resistant populations)

In the mid-west, in many populations of common waterhemp (pigweed species), biotypes differed in susceptibility to ALS-inhibiting herbicides. With recurrent spraying of ALS-inhibiting herbicides, populations shifted from susceptible to highly resistant biotypes. (Note: A concurrent shift in many weed communities also occurred. Other species were controlled with these highly effective herbicides, and the waterhemp numbers increased as a proportion of the entire weed community). Other ALS-resistant pigweed species have also developed in several areas of the United States, including the northeast. Most recently, glyphosate resistant weeds such as horseweed (marestail) and pigweed species are a problem in different regions of the U.S. as a direct result of glyphosate use in herbicide resistant crops.

Production Practices that Influence Shifts

Effectiveness of Weed Management Program

The effectiveness of the overall weed management program will be the single most important factor affecting weed shifts. If the program is not managing all weed species, species that are not effectively controlled will increase. Farmer diligence and ability to make adjustments in a weed management program will directly influence the occurrence of weed shifts.

Speed of Weed Shift Occurrence

The speed at which weed shifts occur will depend on various factors. The ability to prevent the introduction of new weeds to an area, the buffering capacity of the soil seed bank, weed characteristics described previously, and the farmer's ability to quickly adjust a weed management strategy when a shift is first observed will all influence the speed at which a shift occurs - or if a shift occurs at all. With poor management, a field could shift predominant weed species very quickly. For instance, a single application of Roundup early in a field with late-emerging large or smooth crabgrass could result in a severe problem with these annual grasses or other similar late-emerging weeds.

Interface Areas

A cropped field will likely include places where no herbicide is applied (skips) or where reduced rates are applied (field edges, places with a lack of overlap). These areas may or may not influence weed shifts. In some cases, these areas may allow genotypes that were not subjected to the selection pressure to survive and reproduce. However, these areas may be so small that they have a negligible effect on overall field effects.

Shifting of Weed Flora in Rice-Wheat Cropping System

Weed flora of this intensified cropping system is dominant with the annual weeds. Now a day's weed management has been associated with second generation problems. In the early 1990's the menace of *Phalaris minor* has increased in rice-wheat cropping system and has threatened the productivity of wheat crop. Continuous use of urea based herbicides for more than a decade has resulted in the evolution of resistant biotypes of *Phalaris minor* now the situation is under control but the problem of cross resistance is emerging. Three new herbicides (sulfosulfuron, clodinafop and fenoxaprop) have been recommended for the control of isoproturon-resistant biotypes of *Phalaris minor* in rice-wheat system. Now the cases of cross-resistance against these herbicides particularly clodinafop after its continuous use have been reported from the farmers in the rice-wheat belt. Rice crop management system supports the survival of *P. minor* seed in rice-wheat system. New emerging weeds like *Sphenoclea zeylanica* in rice and *Rumex retroflexus* and *Malva parviflora* in wheat may be the likely menace of future in rice-wheat system.

Influence of Long term Herbicide use on Weed Shifts in Maize

Weed species	Year	
	2013	2014
Achillea millefolium	–	+
Artemisia vulgaris	+	+
Amarenthus retroflex	+	+
Cirsium arvensis	–	+
Gallium aparine	+	–
Lactu caserriola	–	+
Daucus carota	–	+

Prevention of Weed Shifts

Weed shifts can be prevented by following methods.

Frequent Monitoring for Escapes

It is difficult to detect an emerging weed shift or weed resistance problem if fields are not frequently monitored for weeds that escapes current weed management practices. Identification and frequent monitoring can detect problem weeds early and guide management practices, including herbicide selection, rate and timing.

Herbicide Rate and Timing

In weedmanagement programme the grower must be sure to use the proper herbicide rate for the particular weeds species as they may be sometimes tolerant to lower doses. And also the time of application of the herbicide dose is important it treat the weeds when they are small, because after crossing certain stage they may be tolerant to that particular herbicide or dosage.

Crop Rotation

One of the most effective practices for preventing weedshifts and weed resistance is crop rotation, which allows growers to modify selection pressure imposed on weeds. Crops differ in their ability to compete with weeds; some weeds are a problem in some crops, while they are less problematic in others. Rotation therefore would not favor any particular weed spectrum. Crop rotation also allows the use of different weed control practices, such as cultivation and application of herbicides with different sites of action. As a result, no single weed spices or bio-type should become dominant.

Agronomic Practices

In addition to crop rotation, several management practices may have an impact on the selection of problem weed populations. If problem weeds germinate at a specific time of year, crop seeding date can be shifted to avoid these weed populations. Delaying irrigation after can reduce germination of certain summer annual weeds. However, this practice only works on some soil types and water stress resistant crops only. Harvest management can, assist in eliminating or suppressing problem weed populations in some cases, but harvest must occur before weed seed production to prevent weed proliferation.

Rotation of Herbicides

Weed shifts occur because herbicides are not equally effective against all weed species and herbicides differ greatly in the weed spectrum they control. A weed species that is not controlled will survive and increase in density following repeated use of one herbicide. Therefore herbicides rotation is recommended. Rotation of herbicides reduces weed shifts, provided the rotational herbicide is highly effective against the weed species that is not controlled with the primary herbicide. The grower should rotate to an herbicide with a complimentary spectrum of weed control, along with a different mechanism of action and therefore a different herbicide binding site. Weed susceptibility charts are useful to help develop an effective herbicide binding site and herbicide rotation scheme.

Rotating herbicides is also an effective strategy for resistance management. Within a weed species there are different bio-types, each with its own genetic makeup, enabling some of them to survive a particular herbicide application. The susceptible weeds in a population are killed, while the resistant ones survives, set seed, and increase over time. Using an effective herbicide with a different mode of action from the one to which the weeds are resistant, however, controls both the susceptible and resistant bio-types. This prevents reproduction and slows the spread of the resistant bio-type.

4.1

Economic Analysis of Weed Management Practices

Weed Management

The productivity of crops much depends on intercultural and weeding operation. The efficient management practices ultimately increases the yield level. The productivity is much dependent on various attributes starting from the variety, climate including human resource management and other factors known well to the farmers. It is estimated that nearly 30-50 per cent of the crop is lost due to weed menace in the agricultural fields. The farmers use various methods of weed control like cultural, mechanical, chemical and manual means. Encompassing all these control measures is known as integrated weed management. The conceptual model of integrated weed management is presented as Figure 4.1.

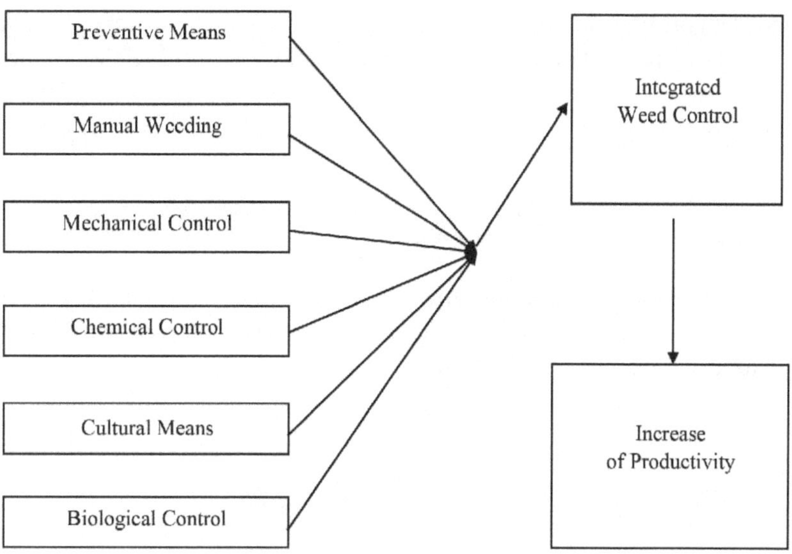

Fig. 4.1. Conceptual Model of IWM

Methods of Weed Management

1. Preventive
2. Manual Weeding
3. Mechanical
4. Chemical
5. Cultural
6. Biological

Encompassing all the above said means of weed management is known as Integrated Weed Management (IWM). It is like any other integrated management practices like pests and diseases management (IPM), nutrient management (INM) and water management IWM). Now it is needed to know something about the economics of management practices as it is the most important factor to decide which management practice is the best one among the available practices. It is nothing but the economic analysis of the various management practices. In other words it is the profitability of adopting a particular management practice.

Economic Analysis

Economic analysis is the analysis of profit or loss accounts in a farm or a firm. Profit is calculated as the difference between the expenditure made and income realized. The profitability of various methods of weed management practices is to be analysed here. There are number of ways to examine the economic viability of the various weed management practices and they are as follows.

1. Farm budgeting
2. Working out the profit/loss
3. Working out the benefit cost ratio

Before carrying out the economic analysis, the method of estimating the cost of cultivation must be known. An illustration is being provided below for knowing the calculation of cost of cultivation of crop.

Table 4.1: Cost of cultivation of Onion (Rs./ha)

S.No.	Particulars Farms	Herbicide Applied weeded Farms	Manually
1.	Labour (weeding)	11143	19730
2.	Labour (Except Weeding)	37300	38250
3.	Machine power	3960	4840
4.	Seed (bulbs)	31350	32330
5.	Organic manures	4900	5580
6.	Inorganic fertilizers	4700	4733
7.	Cost of labour and other inputs	93353	105463
8.	PP chemicals (including herbicides)	3127	1827
9.	Transportation	1280	1570
10.	Interest on working capital	3400	3800
11.	Total cost/Gross Expenditure	101160	112660
12.	Yield (kgs/ha)	20395	17525
13.	Price of Onion (Rs/kg)	15.00	15.00
14.	Value of Main Product	305925	262875
15.	Net income	204765	150215
16.	Net income Rs./Kg	10.04	8.57
17.	BC Ratio	3.02	2.33

Budgeting

It may be defined as a detailed physical and financial statement of a farm plan or of a change in farm plan over a certain period of time. In other words, the expression of farm plan in monetary terms through the estimation of receipts, expenses and profit is called farm budgeting.

Types of Farm Budgeting

The following are the different types of farm budgeting techniques:

a) Partial Budgeting

b) Enterprise Budgeting

c) Cash flow Budgeting

d) Complete Budgeting

a) Partial Budgeting

This refers to estimating the outcome or returns for a part of the business, i.e., one or few activities. A partial budget is used to calculate the expected change in profit for a proposed change in the farm business. A partial budget contains only those income and expense items, which will change, if the proposed modification in the farm plan is implemented.

Limitations of Partial Budgeting Technique

1. Partial change does not always provide a complete solution.
2. The results of partial budgets are subject to variations in output-input prices, availability of resources and variations due to soil type, soil fertility etc.

E.g. Partial budgeting technique was used to find out the usefulness of the chemical method of weed control. In order to assess the benefits in monetary terms, the cost of cultivation was estimated for both the category of farmers' viz., adopters of chemical method of weed control and otherwise. The results of the partial budgeting analysis is given below.

Table 4.2: Partial Budget showing change between users and non-users of Herbicide in

	Debit (A)			Credit (B)	
I.	**Added Cost**		**I**	**Added return**	
i.	Cost of herbicides	573	i.	Gross return*	7622
ii.	Cost of fertilizers	692			
iii.	Cost of machine hours	752			
II.	**Reduced revenue**	Nil	**II**	**Reduced Cost**	
			i.	Cost on human labour	419
.	A. Total added cost and Reduced return	2017		B. Total added return and reduced cost	8041

*Difference between the gross return of users of Herbicide and Non-users of herbicide

Net Change in Income: B - A = Rs. 6024

It could be seen from the above table that the net change in income was positive at Rs. 6024 per ha. The inference is that the adoption of chemical method of weed management yields positive additional revenue. This has indicated that the technology could well be adopted by the farmers in mitigating their problem of weed management.

b) Enterprise Budgeting

An enterprise budget is an estimate of all income and expenses associated with a specific enterprise and an estimate of its profitability. It includes total operating cost and fixed cost including depreciation and interest on fixed asset.

Table 4.3: Enterprise Budget for Irrigated Ground Nut

(Rs/hectare)

Particulars		Amount
I Returns		
1. Main product		6277
2. By product		524
3. Gross return		6801
II Cost		
1.	Land revenue	21
2.	Seed	1245
3.	Manures and fertilizers	530
4.	Plant protection chemicals	98
5.	Irrigation charges	190
6.	Machine power	206
7.	Bullock power	304
8.	Human labour	1617
9.	Interest on working capital	128
10.	Depreciation on buildings and machineries	125
11.	Interest on Fixed capital	725
	Total cost	**5189**
	Net return	**1612**

c) Cash - Flow Budgeting

It is essential to know about cash flow statement before using the cash flow budgeting. The cash flow statement summarizes the magnitude of cash inflows and outflows over a period of time.

Importance of Cash Flow Statement: It helps to assess:

i) Whether cash would be available in correct quantity at right time

ii) Whether the surplus could be profitably diverted and

iii) Timing and magnitude of borrowings required.

Complete or Whole Farm Budgeting

It is generally used by beginners or by those farmers who want to completely overhaul their existing farm organization and operation. Complete and partial budgeting is mutually complementary, i.e., the partial budgeting should be used at various stages of complete budgeting in order to decide the changes to be effected in the farm organization.

Table 4.4: Complete Budget Showing Projected Income, Expenses and Profit.

Particulars	Amount (Rs)
I. Income from various sources	
i) Cotton	54,000
ii) Paddy	43,000
iii) Sorghum	13,500
iv) Dairy products	40,000
Total income	150,500
II. Total Variable Expenses	74,500
III. Total Fixed Expenses	36,250
Net Farm Income {(I – (II+III)}	**39,750**

Farm Profitability

Finally, it is the desire of any farmer to assess his position in terms of profitability. The farm profitability could be estimated by summing up all the expenditures made in the farm as well as summing up the income realized from all activities in the farm. The farm profit is calculated as below.

Farm Profit = Income Realized (Gross Income)-Expenditure made (Gross Cost)

Benefit Cost Ratio (BC Ratio)

The benefit cost ratio reveals the returns obtained by the farmer for an investment of one rupee. It is the ration between the income realized and the expenditures made during the process of any particular crop or the farm as a whole.

$$\textbf{BC Ratio} = \frac{\text{Income realized (Gross Income)}}{\text{Expenditure made (Gross Cost)}}$$

BC Ratio

Problem

Kindly calculate the **cost of cultivation, profit and BC ratio** of a farmer cultivating tomato in one hectare of land. He has incurred the costs and realized an income as below.

Table 4.5: Costs and realized income of Tomato

Particulars	Amount (Rs.)
Male Wages	3559
Female Wages	9384
Animal hire charges	355
Machine hire charges	2412
Seed Value	2833
FYM Value	6100
Fertilizer value	4977
MN Value	360
Herbicide value	1090
PP Chemicals value	1682
Other expenditure	4254
Income realized	92245

Economics of Weed Management Practice in Cotton

The cost of cultivation, gross return and net return under different pre and post-sowing weed management practices on cotton were worked out and benefit cost ratio was calculated (Table 4.6) (Veeramani., 2004).

The pre-sowing weed management practices of SSB by paraquat application (S_3) recorded a maximum net return of Rs. 14,570 ha^{-1} which was followed by SSB by slight hoeing (Rs. 11298 ha^{-1}). The minimum net return (Rs. 10724 ha^{-1}) was obtained in normal sowing (S_1). In the case of post-sowing weed management practices, manual weeding (20 DAS) fb glyphosate 1.5 kg ha^{-1} (W_4) registered higher net return (Rs. 21,742 ha^{-1}) than manual weeding (20 DAS) fb glyphosate 1.0 kg ha^{-1} (W_5) practice (Rs. 17,505 ha^{-1}) and farmer's practice practice of manual weeding twice (W_6) treatment (Rs.16,716 ha^{-1}).The unweeded control (W_7) recorded the lowest net return of Rs. 318 ha^{-1}.

Higher benefit cost ratio was obtained with SSB by paraquat application (S_3) practice (2.06) followed by SSB by slight hoeing (S_2) practice (1.82). The normal sowing (S_1) was not economically viable as it recorded the lowest benefit cost ratio (1.81). In the case of post-sowing weed management practices, manual weeding (20 DAS) fb glyphosate 1.5 kg ha^{-1} (W_4) registered higher benefit cost ratio (2.36).

Table 4.6 Effect of pre and post sowing weed management practices on economics of cotton (Veeramani, 2004)

Treatments	Cost of cultivation (Rs ha^{-1})	Cost of weeding (Rs ha^{-1})	Total cost of cultivation (Rs ha^{-1})	Seed cotton yield (kg ha^{-1})	Gross return (Rs ha^{-1})	Net return (Rs ha^{-1})	Net gain or loss over unweeded control (Rs ha^{-1})	Benefit cost ratio
Pre-sowing weed management practices								
S_1-Normal sowing	13124	-	13124	1084	23848	10724	10406	1.82
S_2-SSB by slight hoeing	13124	900	14024	1151	25322	11298	10980	1.81
S_3-SSB by paraquat	13124	686	13180	1290	28380	14570	14252	2.06
Post-sowing weed management practices								
W_1-Fluchloralin 1.0 kg/ha-MW	13124	3080	16204	865	19030	2826	2508	1.17
W_2-Fluchloralin 1.0 kg/ha-glyphosate 1.5 kg/ha	13124	2300	15424	1180	25960	10536	10218	1.50
W_3- Fluchloralin 1.0 kg/ha-glyphosate 1.0 kg/ha	13124	2007	15131	1033	22726	7595	7277	1.50
W_4- MW-glyphosate 1.5 kg/ha	13124	2820	15944	1713	37686	21742	21424	2.36
W_5- MW-glyphosate 1.0 kg/ha	13124	2527	15651	1598	35156	17505	19187	2.25
W_6-MW (20 and 40 DAS)	13124	3600	16724	1520	33440	16716	16398	2.00
W_7-Unweeded	13124	-	13124	611	13442	318	-	1.02

Cost of herbicides (Rs lit^{-1});
Paraquat : 340.00
Fluchloralin : 480.00
Glyphosate : 240.00
MW-Manual Weeding

Hire charges for tractor: Rs 125 hr^{-1};
Cost of labour (Rs. day^{-1})
A type: 72.00
B type: 72.00
SSB-Stale seed bed

Cost of fertilizers (Rs kg^{-1});
Urea : 4.90
Super phosphate : 3.20
Muriate of potash : 4.50

Cost of seed cotton: Rs. 65.00 kg^{-1}
Sale price of cotton: Rs. 2,200 qtl^{-1}

Colour Plates

Chapter 1: Introduction to Weeds

Fig. 1.1.1
Commelina benghalensis

Fig. 1.1.2
Boerhavia erecta

Fig. 1.1.3
Chenopodium album

Fig. 1.1.4 *Sonchus arvensis*

Fig. 1.1.5 *Phleum pratense*

Fig. 1.1.6 *Sorghum halapense*

Fig. 1.1.7 *Cynodon dactylon*

Fig. 1.1.8 *Convolvulus arvensis*

Fig. 1.1.9 *Cyperus rotundus*

Fig. 1.1.10 *Ammania baccifera*

Fig. 1.1.11 *Eclipta alba*

Fig. 1.1.12
Trianthema portulacastrum

Fig. 1.1.13
Digera arvensis

Fig. 1.1.14 *Tribulus terrestris*

Fig. 1.1.15 *Argemone mexicana*

Fig. 1.1.16. *Aristolochia bracteata*

Fig. 1.1.17. *Leucas aspera*

Fig. 1.1.18. *Lantana camara*

Fig. 1.1.19 *Spergula arvensis*

(Grass)
Fig. 1.1.20 *Cynodon dactylon*

(Sedge)
Fig. 1.1.21 *Cyperus rotundus*

(Broad leaf weed)
Fig. 1.1.22 *Tridax procumbens*

Fig. 1.1.23 *Gynandropsis pentaphylla*

Fig. 1.1.24 *Calotropis gigantea*

Fig. 1.1.25 *Phalaris minor*

Fig. 1.1.26 *Indigofera enneaphylla*

Fig. 1.1.27 *Acalypha indica*

Fig. 1.1.28 *Abutilon indicum*

Fig. 1.1.29 *Parthenium hysterophorus*

Fig. 1.1.30 *Acanthospermum hispidum*

(Monocot weed)
Fig. 1.1.31: *Echinochloa colona*

(Dicot weed)
Fig. 1.1.32: *Crotalaria verucosa*

Fig. 1.1.34 *Taraxacum* sp.

Fig. 1.1.33 *Rumex acetosella*

(Woody weed)
Fig. 1.1.35 *Prosopis juliflora*

(Semi-woody weed)
Fig. 1.1.36 *Croton sparsiflorus*

(Herbaceous weed)
Fig. 1.1.37 *Amaranthus viridis*

Fig. 1.1.38

Fig. 1.1.39

Fig. 1.1.40

Fig. 1.1. 41 *Orabanche cernua* on Tobacco

Fig. 1.1.42: *Striga lutea* on sugarcane

Fig. 1.1.43 *Cuscuta chinensis*

Fig. 1.1.44 *Loranthus longiflorus* on mango

Fig. 1.1.45 : *Eichhornia crassipes*

Fig. 1.1.46 *Salvinia* sp.

Fig. 1.1.47 *Utricularia stellaris*

Fig. 1.1.48 *Ceratophyllum demersum*

Fig. 1.1.49 *Nelumbium speciosum*

Fig. 1.1.51 *Typha* sp.

Fig. 1.1.52 *Polygonum* sp.

Chapter 2: Weed Biology

Evening prime rose

Wild carrot

Fern plant

Fern spores

Canada thistle

Cyperus esculentus– tubers

Tridax procumbens

Pappus seed

Curly dock

Plant

Seed

Mature seed

Xanthium strumarium -hook

*Cenchrus*spp - stiff hair

Tribulusterrestris - spines

Avena fatua

Phalaris minor

Chapter 2.1: Weed Control Methods

Straw mulch

Polythene mulch

Chapter 2.5: Herbicide Application Techniques

Fig.2.5.12 Aerial application of herbicide

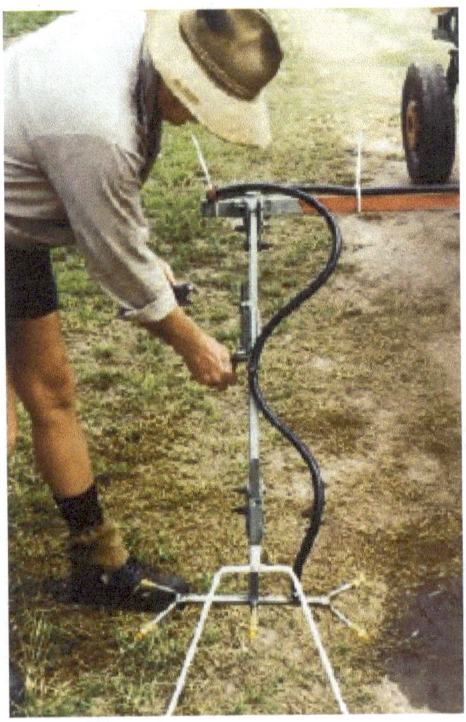

Fig.2.5.13 Modern applicant to apply herbicides

References

Ahuja, K.N. and Yaduraju.1989. Integrated control of weeds in wheat with special reference to *Phalaris minor*. *Indian J. Weed Sci*. 30 (3&4): 141-144.

Anonymous, 1992. Proceedings of workshop on 'Management of *Parthenium*- A critical appraisal' held on October 22nd , 1992 at I.A.T., Bangalore.

Ashton,F.Y. & Monaco,T.J. 1991. Weed Science Principles and Practices", New York. John Wiley & Sons, Inc.

Bhan,V.M. 1987. Paper presented in PAK –INDO -US weed control workshop-cum-first annual conference, Pakistan Society of Weed Sci. Islamabad, March 11-14, 1987 pp.76-89.

Banting,J.D. 1970. Effect of diallate and triallate on wild oat and wheat cells. *Weed Sci.* 18 (1): 30-84.

Barrentine, W.L. and Warren, G.R.1971. Differential phytotoxicity of trifluralin and nitralin. *Weed Sci.* 19 (1): 31-37.

Bhowmik, P.C. and Doll, J.D.1992. Corn and Soybean response to allelopathic effects of weed and crop residues. *Agron*. J. 74:601-606.

Blossey B and Notzold R.1995. Evolution of increased competitive ability in invasive on-indigenous plants: a hypothesis. J Ecol 83:887–889.

Boldt, P.F. and Putnam, A.R.1981. Selective mechanisms for the foliar application of dichlofop-methyl –II-metabolism. *Weed Sci.* 29 (2) : 237-241.

Center, T.D.1982. The water-hyacinth weevils. *Aquatics* 4 (2):8.

Copping, L.G. and Hewitt H.G. eds.1998. Chemistry and mode of action of crop protection agents. The Royal Soc. of Chemistry, Cambridge, U.K, pp.17-45.

Corn, J.S. and Farris, M.L.1983. Seed viability and dormancy of 17 weed species after 21 months in Alaska. *Weed Sci.* 35(4):524-529.

Crutchfield, D.A., Wicks, G.A. and Buruside, D.C.1982. Role of wheat straw mulch on weed growth in maize. *Weed Sci.*34:110-114.

Dekker, J. and Duke, S.O. 1995. Herbicide resistant field crops. *Adv. In Agron* 54 : 69-113.

Devine, M.D., Duke, S and Fedtke, C.1993. Physiology of herbicide action. Prentice Hall, Englewood, Cliffs, Newzealand.

Dixon, G.A. and Stoller, E.W.1982. Differential toxicity of metolachlor in corn and yellow nutsedge. *Weed Sci.* 30:225-230.

Durborow R, Tucker C, Gomelsky B, Onders R and Mims S .2008. Aquatic weed control in ponds. Kentucky State University Land Grant Program, 24 p.

Frear, D.S. and Still, G.G.1968. The metabolism of 3, 4-dichloropropionanlide in plants. Partial purification and properties of an arylacylamidase from rice. *Phytochem* 7: 913-920.

Frear, D.S., Swanson, H.R. and Tanaka, F.S. 1969. N demethylation of substituted 3-(phenyl) -1-methylureas: Isolation and characterization of a microsomal mixed furation oxidase from cotton. *Phytochemistry* 8: 2157-2169.

Fuerst, E.P. and Norman, M.A. 1991. Interactions of herbicides with photosynthetic electron transport. *Weed Sci.* 39: 458-464.

Gill, H.S. and Walia, U.S.1987. Influence of ponding water on the bio efficacy of herbicides in transplanted rice grown on light textured soil *Oryza* 24: 146-149.

Graf, B and Hill, J.E. 1992. Modelling the competition for light and nitrogen between rice and *Echinochloa crusgalli*. Agri. Systems 40 (94): 345-359.

Gupta, O.P. *et al.* 1979. Evaluation of response of potato to chemical control of early weed infestation. IAAS. J. 1 (4): 139-143.

Gupta, O.P.1998. "Modern weed management". Published by Mrs. Saraswati for Agro Botanica, U.E-176, J.N. Vyas Nagar, Bikaner, India.

Gutterman, Y. 1985. Flowering, seed development and he influences during seed maturation on seed germination of annual weeds. In: *Weed Physiology, Reproduction and Ecophysiology* (Ed. S.O. Duke,pp 2-23.

Hager, A.G. and Nordby, D. 2007. Herbicide persistence and how to test for residues in soils. Illinois agricultural pest management hand book. pp. 345-350.

Hess, F.D. and Bayer, D.1977. Binding of the herbicide trifluralin to chlamydomenas flagella tubulin. *J Cell Sci.* 24: 351-360.

Holt, J.S., Holtum, J.A.M. and Powles, S.B.1993a. Mechanical and agronomic aspects of herbicide. *Ann. Rev. Plant Physio. Plant Mol. Biol.* 44: 203-209.

Hongo, A.1994. Longevity of buried weeds of three *Rumex spps* over a fourteen year period. *Weed Sci.* (Tokyo) 39(1): 52-53.

Jaya Kumar, R. and Sankaran, S. 1988. Quantitative determination of butachlor residues in rice and subsequent blackgram. *Biennial Conf. Indian Soc. Weed Sci.*, March 8-9.

Kigel,J., Gibly, A. and Negbi, M.1979. Seed germination in *Amaranthus retroflexus* as affected by the photoperiod and the age during flower induction of the parent plants. *Journal of Experimental Botany* 30:997.

Kigel, J., Ofir, M. and Koiler, D. 1977. Control of the germination responses of *Amaranthus retroflexus* L. seeds by their parental photothermal environment. *Journal of Experimental Botany* 28:1125.

Kiviban, A and Bavdurski, R.S.1973.The ninety year period for Dr. Beals' seed viability experiment. Am. J. Bot. 60: 140-145.

Kolar, J.S. and Mehra, S.P.1992. Changing scenario of weed flora in agroecosystems of Punjab. (IN) Changing scenario of Our Environment. (Eds) Dhaliwal, G.S. *et al.*, P.A.U., Ludhiana (India): 252-262.

Kulshrestha, G.1982. Hydrolysis of isoproturon in aquous medium and its persistence in soils and plants. *Indian J Weed Sci.* 14: 96-102.

Lichtner, F.T.1995. Ethametsulfuron methyl metabolism and crop selectivity in spring oilseed rape. *Pesticide Biochem & Physiol.* 52 (1): 12-24.

Mahajan, G., Brar, L.S. and Walia, U.S. 2000. Paper presented in *International Conference on Managing Natural Resources for Sustainable Agricultural Production in the 21st Century* held at New Delhi from Feb. 14-18,2000 pp.1289-1291.

Malik, R.K., Gill, G and Hobbs, P.R.1998. Herbicide resistance –A major issue for sustaining wheat productivity in rice wheat cropping system in Indo-Gangetic plains. In: Rice wheat Consortium for Indo –Delvi, pp.1-32.

Mani,V.S. 1975.Nutrient drain by weed growth in crop fields. *Fert. News* 20 (3): 21-27.

Matsumoto, H.1999. Herbicide mode of action: Recent progress and new challenges in Proc. 7th Asian Pacific Weed Sci. Soc. Conf. Nov.22-27, 1999, Bangkok, Thailand, pp. 39-44.

Mishra, J.S.1997. Critical period of weed competition and losses due to weeds in major field crops. *Farmer and Parliament* 33 (6): 19-20.

Prendeville, G.N., Eshel, Y., James, C.S., Warren, G.F. and Schreilber, M.M.1968. Movement and metabolism of CIPC in resistant and susceptible species. *Weed Sci.* 16: 432-435.

Radix, P., Belleville, L., Eygven, J. and Chevallier, P.1994. Pesticides pollution in several drinking water networks in the south eastern parts of France. Fresenius Environmental Bull 1: 536-541.

Rahman, A., James, T.K. and Gunther, P. 1993. Bioassays of soil applied herbicides. In: Proc. Int.Symp. Indian Society of Weed Science, Hissar, Nov. 18-20, 1993. I: 95-106.

Randhawa, S.K. and Sandhu, K.S.1992. Persistence of isoproturon and methabenzthiazuron applied for weed control in onion (*Allium cepa*). *Indian J. Agron.* 37: 620-622.

Rao, V.S.2000.Principles of Weed Science. Oxford and IBH Publishing Co., New Delhi.

Robert M.Durborow. 2014. Management of Aquatic Weeds. In: Bhagirath S. Chauhan, Gulshan Mahajan (eds). Recent Advances in Weed Management. Springer, New York, pp 281-313.

Saini, J.P., *et al.* .1993. Efficacy of atrazine at different moisture levels for weed management in sugarcane under mid-hill conditions of Himachal Pradesh. *Ind. J. Weed Sci.* 25 (3&4):49-55.

Scheunert, I. 1992. In: Chemistry of Plant Protection. Vol. 8 (Ebing, Weed.) Springer, Berlin/ Heidelberg, pp. 23-75.

Shahi, H.N.1978. Competetive effects of *Chenopodium album* for soil moisture and nutrients in wheat. *Ind. Pest Control* 20 (4) :14-16.

Singh, *M. et al.* 1978. Crop competition studies in mungbeans. *Ind. J. Agron.* 23 (4): 377-378.

Singh, T. *et al.* 1993. Weed Flora of wheat fields in the Punjab. Ind. J. Weed Sci. 25: (1&2) 17-21.

Singh, H. and Kolar, J.S.1994. Crop- weed competition studies in soybean (*Glycine max* L. Merril). *Ind. J. Weed Sci.* 26 (1&2): 83-86.

Stalder, L. and Pestemer, W. 1980. Availability to plants of herbicide residue in soil Part I: A rapid method for estimating potentially available residues of herbicides. *Weed Res.* 20: 341-347.

Subramaniam, S., Mohamed Ali,A., Muthukrishnan, P., Jaya Kumar, R., Chandrasekaran, B and S.Vallal Kannan.2011. All About Weed Control. Kalyani Publishers, New Delhi, India

Suzuki, M.1994. Observations on germination ability of upland weeds burried into soil at a 30cm depth during 25 years. *Weed Res.* (Tokyo) 39(1): 34-39.

Swain, T. and Williams, C.A.1977. Heliantheae-chemical review. In: The Biology and Chemistry of the of the Compositae. Vol. II V.H. Hay Wood, J.B.Harborne and B.L. Turner (eds) Academic Press, London. pp.673-697.

Tomers, G.H.N and Subba Rao P.V.1992. Impact of the pan-tropical weed. *Parthenium hysterophorus* L. on human affairs. proc. 1st Int. Weed Control Cong. Melbourne I: 134-138.

Veeramani, A.2004. Effect of Pre and post sowing weed management in Summer irrigated Cotton. Ph.D. Thesis, Tamil Nadu Agric. Univ., Coimbatore.

Virk, H.K, H.S. Brar and U.S.Walia. 2003. Competitive ability of wheat cultivars sown on different dates with Little seed canary grass (*Phalaris minor* Retz.). *Indian J. Weed Sci.*35 (1&2):21-23.

Walia, U.S.2003.Weed Management. Kalyani Publishers, New Delhi.

Walia, U.S., Brar, L.S. and Dhaliwal, B.K.1997. Resistance to isoproturon in *Phalaris minor* Retz. in Punjab. *Plant Protec. Quarterly* 12(93): 138-140.

Walia, U.S. and Gill, H.S. 1985.Influence of Nitrogen and Substituted Urea Herbicides on the uptake of N, P and K by *Pahalaris minor* Retz. and wheat. *Indian J. Weed Sci.* 17 (1):12-17.

Werker,E.1980. Seed dormancy as explained by the anatomy of embryo envelopes. *Israel Journal of Botany* 29:22

Wilcove DS, Rothstein D, Dubow J, Phillips A and Losos E.1998. Quantifying threats to imperilled species in the United States. BioScience 48: 607–615.

Yenish, J.P. *et. al.*1992. Effects of tillage on vertical distribution and viability of weed seeds in soil. *Weed Sci.,* 40(3):429-433.

FSC
www.fsc.org
MIX
Papier | Fördert
gute Waldnutzung
FSC® C083411

Zeitfracht Medien GmbH
Ferdinand-Jühlke-Straße 7
99095 Erfurt, Deutschland
produktsicherheit@kolibri360.de